COMMUNICATION AND BEHAVIOR

AN INTERDISCIPLINARY SERIES

Under the Editorship of ***Duane M. Rumbaugh,***

Georgia State University and Yerkes Regional Primate Research Center of Emory University

DUANE M. RUMBAUGH (ED.), LANGUAGE LEARNING BY A CHIMPANZEE: THE LANA PROJECT, 1977

ROBERTA L. HALL AND HENRY S. SHARP (EDS.), WOLF AND MAN: EVOLUTION IN PARALLEL

HORST D. STEKLIS AND MICHAEL J. RALEIGH (EDS.), NEUROBIOLOGY OF SOCIAL COMMUNICATION IN PRIMATES: AN EVOLUTIONARY PERSPECTIVE

P. CHARLES-DOMINIQUE, H. M. COOPER, A. HLADIK, C. M. HLADIK, E. PAGES, G. F. PARIENTE, A. PETTER-ROUSSEAUX, J. J. PETTER, AND A. SCHILLING NOCTURNAL MALAGASY PRIMATES: ECOLOGY, PHYSIOLOGY, AND BEHAVIOR

NOCTURNAL MALAGASY PRIMATES

Ecology, Physiology, and Behavior

P. Charles-Dominique, H. M. Cooper, A. Hladik, C. M. Hladik, E. Pages, G. F. Pariente, A. Petter-Rousseaux, A. Schilling

Centre National de la Recherche Scientifique
Brunoy, France

J. J. Petter

Muséum National d'Histoire Naturelle
Brunoy, France

1980

ACADEMIC PRESS

A Subsidiary of Harcourt Brace Jovanovich, Publishers

New York London Toronto Sydney San Francisco

ACADEMIC PRESS, INC.
111 Fifth Avenue, New York, New York 10003

United Kingdom Edition published by
ACADEMIC PRESS, INC. (LONDON) LTD.
24/28 Oval Road, London NW1 7DX

Library of Congress Cataloging in Publication Data

Charles-Dominique, Pierre.
Nocturnal Malagasy primates.

(Communication and behavior)
Includes bibliographies and index.
1. Lemurs--Behavior. 2. Lemurs--Ecology. 3. Mammals--Behavior. 4. Mammals--Ecology. 5. Mammals--Madagascar. I. Title. II. Series.
QL737.P95C45 599.8'1 79-6799
ISBN 0-12-169350-3

PRINTED IN THE UNITED STATES OF AMERICA

80 81 82 83 9 8 7 6 5 4 3 2 1

In the memory of Georges F. Pariente
(1937–1976)

Georges F. Pariente measuring the spectral composition of light at dusk, near Marosalaza Forest, in 1974.

Contents

PART II STUDIES IN SIMULATED CLIMATIC CONDITIONS

Preface

The Malagasy fauna is a remarkable example of adaptive radiation in terms of morphological, physiological, as well as behavioral characters evolved in response to diverse environmental conditions. In this respect, the prosimian species inhabiting the dry deciduous forest of Marosalaza, near Morondava, on the west coast of Madagascar, provide a particularly informative case study. Although during the rainy season the forest production is of the same order of magnitude as in other tropical ecosystems, the dry season is typified by an extreme scarcity of food resources. These marked contrasts in climate and production have resulted in the development of distinct and unusual adaptations of the various prosimian populations.

The different studies presented in this volume were undertaken to investigate the "strategies" adopted by these species to cope with the contrasted seasonal conditions. Until recently, most ecological research on the Malagasy lemurs has been focused on diurnal species. In this Morondava program we attempted to define the relationship between ecology and physiology in the lesser known nocturnal species of Marosalaza forest.

As in the previous primatological studies conducted by researchers of the Brunoy laboratory, the first step consisted of defining forest structure, composition, and production during the annual cycle, in relation to climatic conditions. The pluridisciplinary effort was further facilitated by the diversity of approaches utilized by the participants in this project. Field studies were conducted during seven successive and overlapping field trips, from November 1973 to July 1974, and focused on dietary adaptations, growth and reproductive cycles, activity rhythms, social structures, and interspecific competition in the five sympatric prosimian species.

The first five chapters presented in this volume concern this field work, each author attempting to integrate the observations in his particular area to the results of the entire study, in a general concept of coevolution of the plants and animals as respective parts of the forest ecosystem.

Subsequently, laboratory studies conducted at Brunoy in simulated climatic conditions enabled detailed observations on variations in specific physiological and behavioral cycles over a period of 3 years. The relationships and comparisons between species of annual variations in feeding, body weight, activity, and reproduction were the major concerns of this research, with supplementary investigations of social interactions, marking, and learning abilities. The results of these studies are presented in the last five chapters of this volume.

The particular interest of these individual field and laboratory studies derives from their complementarity, and together they constitute a comprehensive approach of the problem of ecophysiology.

This research, which was initially conducted in the field within the scope of "Man and Biosphere" program of UNESCO, was made possible by the helpful cooperation of all Malagasy authorities involved. We are especially indebted to the "Service des Eaux et Forêts de Madagascar" for the participation of J. P. Abraham who contributed in the unique thorough forest inventory of the west coast of Madagascar presented by A. Hladik. We also appreciated the very kind hospitality of the De Heaulme family at the field station of Analabe.

The recent international cooperation developed in conjunction with Malagasy University will hopefully provide the framework for continued in-depth investigations of the problems concerning the interdependence of ecology and physiology, brought to light in this complete but still preliminary Morondava program.

The laboratory studies were only possible thanks to the careful attention of M. Grange, M. T. Brisset, and H. Goujon in the prosimian breeding colony at Brunoy.

The editorial help of H. M. Cooper greatly facilitated the cooperative editorial effort for the final version of the different chapters in this volume. We are also indebted to Alison Jolly for her valuable comments on the various contributions. Typing of the manuscripts and drawings were done by M. Charles-Dominique, C. Munoz, and M. A. Caubere.

The financial support for this program was provided by the Centre National de la Recherche Scientifique (France), to the Equipe de Recherche sur les Prosimiens, E.R.A No. 070410, and A.T.P. No. 449916 Ecophysiologie attributed to J. J. Petter.

The unfortunate accidental death of Georges F. Pariente occurred shortly after the completion of the field studies. He was partly responsible for the original idea of this pluridisciplinary research. The different participants and F. Parra contributed in presenting his last manuscript included in this volume, which is dedicated to his memory.

P. Charles-Dominique
H. M. Cooper
A. Hladik
C. M. Hladik
E. Pages
A. Petter-Rousseaux
J. J. Petter
A. Schilling

Part I

Field Studies

1 The Dry Forest of the West Coast of Madagascar: Climate, Phenology, and Food Available for Prosimians

A. HLADIK

INTRODUCTION

The importance of the interface between plant and animal populations, as an essential level of investigation in tropical ecosystems, is becoming increasingly apparent (Golley and Medina, 1975; Lamotte and Bourlière, 1978). In this respect, the present study of the overall food resources available for prosimians in the forest of the West Coast of Madagascar not only involves trophic relations, but also many phenomena of coevolution. In order to analyze the possible type of interactions, the first step is a thorough inventory of the biotope in terms of plant production and phenology. Accordingly, this chapter is based on collective data from the different participants of the "Morondava programme," mainly those of J.P. Abraham of the "Service des Eaux et Forêts de Madagascar." The author, who did not join the participants in the field, made analyses of forest composition and production based on:

1. The quantitative study of litter fall collected in sampling baskets at regular intervals, by sorting out and weighing leaves and other parts of different species.

2. The identification of the herbarium specimens collected by the different participants, after the tentative field identification by J.P. Abraham. This thorough survey was conducted at the Herbarium of Paris (Muséum National d'Histoire Naturelle).*

* We are much indebted to J. Bosser who aided in identification of many plant species; thanks are due to J. Léandri (identification of Euphorbiaceae) and to M. A. G. Peltier (identification of Leguminosae).

3. A further review of the entire collection of the herbarium specimens, collected previously in the Morondava area, provided additional information on the seasonal cycles of the most important species and complemented field observations and photographs obtained during the field trips of the present program.

4. Alkaloid tests on dry specimens for information on potential toxicity.

A synthesis of these results is presented, with reference and comparisons to previous ecological surveys on the tropical forests of Africa (A. Hladik, 1974, 1978) and in Asia (Hladik and Hladik, 1972) considered from the viewpoints of food resource utilization by primate populations as well as that of plant adaptive strategies.

THE CLIMATE OF THE WEST COAST OF MADAGASCAR

The forest of Marosalaza, near Beroboka, where the different field studies presented in this volume were undertaken, is located 50 km north of Morondava. Due to its latitude (20° S), the Morondava area is characterized by seasonal variations in day length (minimum: 10 h 52 in June; maximum: 13 h 17 in December). Nevertheless, the total number of hours of sunshine is fairly evenly distributed in terms of monthly average (from 263 hours in February to 306 hours in August); for detailed data on light intensity and spectral composition, see Chapter 5.

If we consider the average distribution of rain throughout Madagascar (Fig. 1A), it appears that Morondava (annual rainfall of 800 mm) is not located in the area which receives the lowest quantity of rain. In fact, the distribution of this rain during the seasonal cycle is the important parameter to analyze: most of the precipitation occurs from December to March followed by a long sharply characterized dry season.

According to the average monthly rainfall and temperature (Fig. 2A), 7 to 8 months can be classified as "dry." There are but few tropical forests inhabited by primates in which such a climatic change occurs. For example, there are only 4 "dry" months (Mueller-Dombois, 1968; Hladik and Hladik, 1972) in the deciduous dry forest of Wilpattu (Sri Lanka). On the African mainland, a similar dry climate is also found in some areas of South Africa where bushbabies have been observed in open woodland and "orchard bush" (Bearder and Doyle, 1974); while towards the sahelian zone, in the savana formation of Fété Olé, with 9 dry months, Patas monkeys are nomadic on very large supplying areas (Bourlière *et al.*, 1978).

If we now consider the records of the different years preceding our field

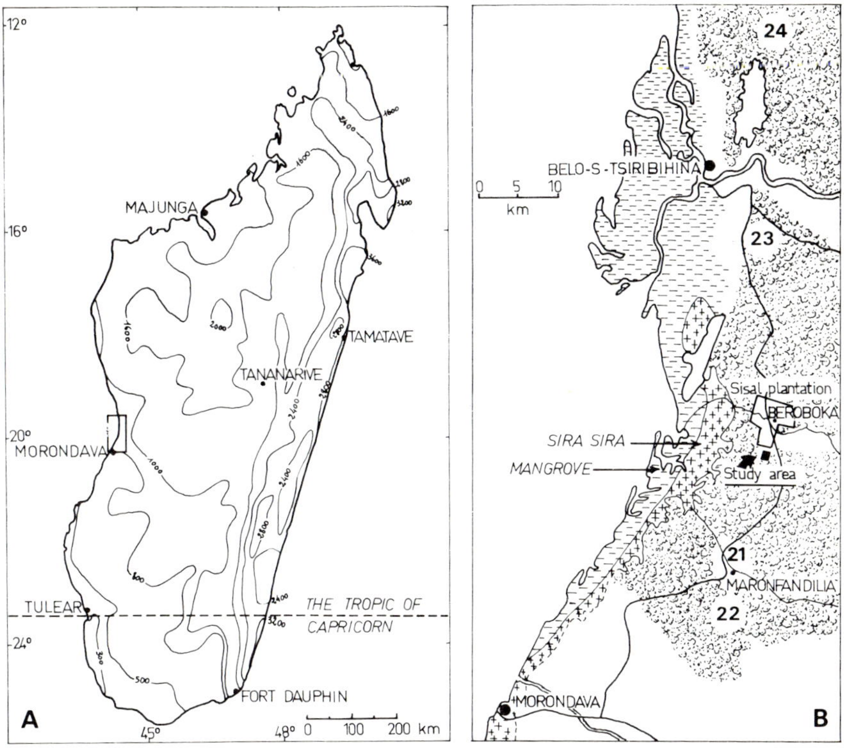

Fig. 1. Location of the study area according to climate and plant formation. A, Distribution of annual rainfall in Madagascar (after Koechlin *et al.,* 1974; Humbert and Cours Darne, 1965). The rectangle near Morondava corresponds to the enlarged map in (B). B, Plant formations in the Morondava area. The study area is located 50 km north of Morondava (black square) in Marosalaza forest. Numbers indicate four different plant surveys of F.A.O. (Anonymous, 1972).

study (taken at the meteorological station of Morondava airport), there are no significant differences in the annual pattern of monthly variation in temperature and humidity (Fig. 2B). In contrast, the pattern of rainfall presents fairly important variations between successive years.

These monthly averages are not precise enough to be correlated with the cycles of vegetation, and thus, during our study period, we collected daily records of rainfall (Fig. 2C) which are more likely to explain the characteristics of the different phases of leafing, flowering, and fruiting of the forest species. During the rainy season, heavy and/or light rains occur at short and regular intervals, while in April/May a few heavy precipitations and storms may irregularly occur.

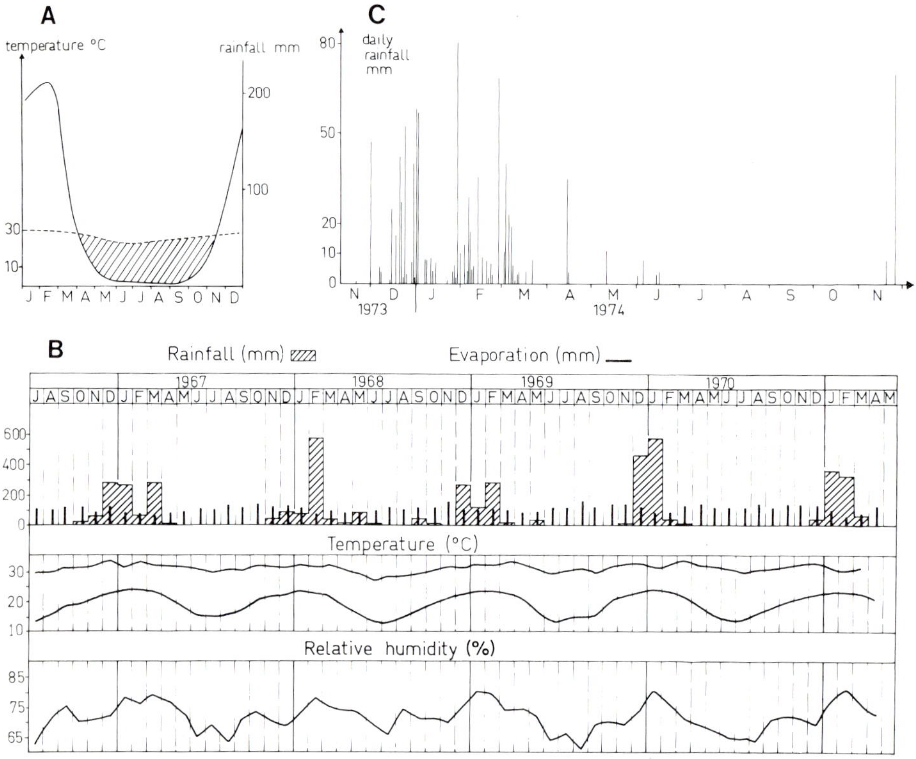

Fig. 2. Meteorological data of the Morondava area. A, The dry months are shown by the shaded area determined by average monthly rainfall (solid line) and average monthly temperature (broken line) (after Legris and Blasco, 1965). Measures were taken at Morondava airport. B, Variations of monthly rainfall and evaporation, temperature (minimum and maximum), and relative humidity, from 1966 to 1971 (after Gachet, 1971). Measures were taken at Morondava airport. C, Daily records of rainfall in the study area (Marosalaza forest) during the study period.

The resulting microclimate inside the forest is illustrated by two patterns characterizing the rainy season (Fig. 3A) and the dry season (Fig. 3B). Temperature and humidity are characterized by very large regular diurnal and nocturnal variations during the dry season. Maximum temperature differences reach 19°C, resulting in important condensation at night. During the rainy season, nights are cold and humidity is constant (around 100%). Nevertheless, due to the buffering effect of the forest, environmental conditions are not different between the dry and the rainy seasons at night when the nocturnal prosimians are active. During the daytime, the condi-

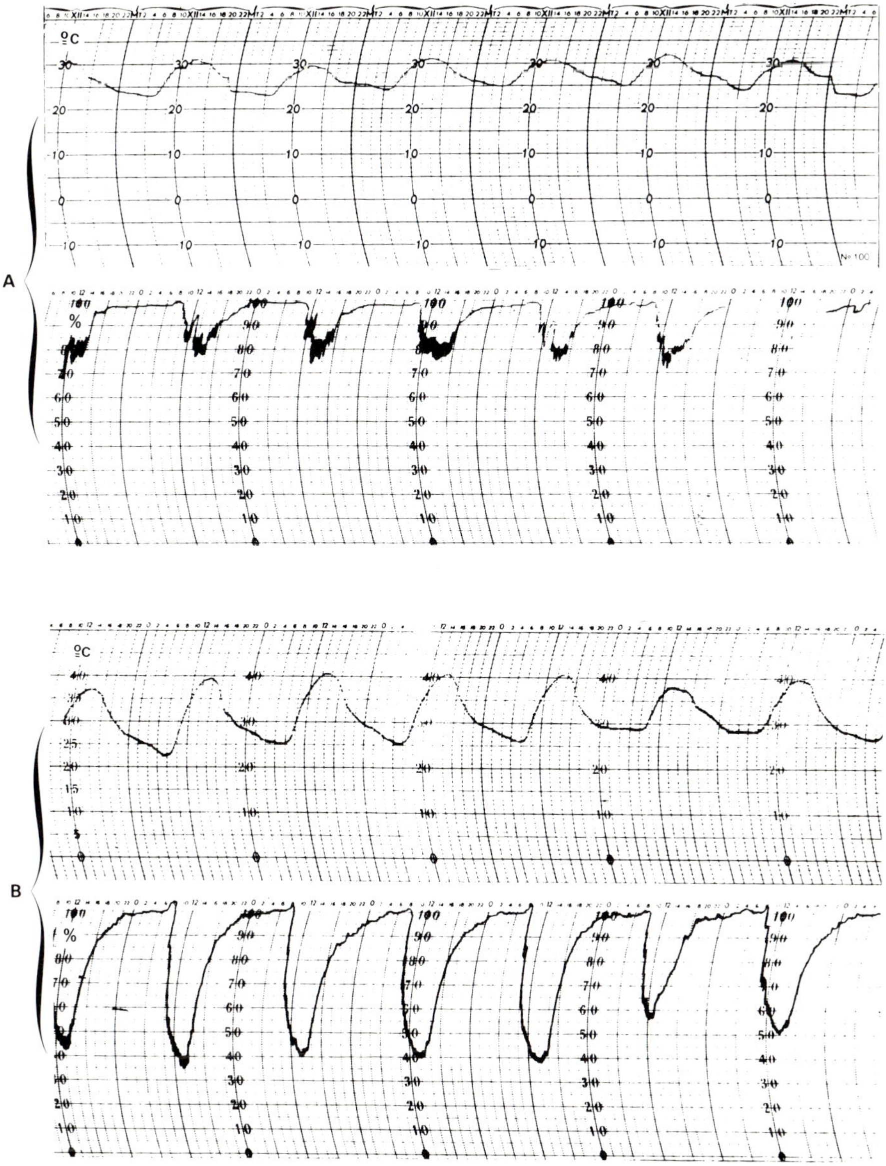

Fig. 3. Variations of temperature (°C) and relative humidity (%) at ground level in the interior of the Marosalaza forest. A, A typical sample (February 25 to March 3, 1974) during the rainy season (austral summer) with relatively low temperature and high humidity due to frequent rain. B, A typical sample (June 16 to June 24, 1974) during the dry season (austral winter). Due to the buffering effect of the forest, humidity remains high at night.

tions in the hollow trunks where these prosimians are found, are partly "buffered" and more similar to nocturnal conditions (cf. Charles-Dominique and Hladik, 1971).

PLANT FORMATIONS IN THE MORONDAVA AREA

The dry forest of Madagascar was proposed at the Yangambi Conference (Aubreville, 1957) as being the typical example of the plant formation called "closed deciduous forest." In the Morondava area, this plant formation presents some peculiar characters (Guillaumet and Koechlin, 1971), the most apparent being the majestic Baobabs which, in terms of biomass, are one of the major components of the forest. The definition given by Perrier de la Bâthie (1921) was "forêt sèche sur sol arenacé du Pliocène" (dry forest on Pliocene sand), while Humbert and Cours Darne (1965) included the plant association in the "série à *Dalbergia-Commiphora-Hildegardia*."

This forest extends along the west coast of Madagascar, as a long strip of about 50 km in width, from the Mangoky River to the Betsiboka River. Like most tropical forests, the structure and the composition are not homogeneous. Included are large parts of alluvial forest bordering rivers and temporary streams, similar in aspect to the gallery forest of the South (Jolly, 1966). There are also temporary ponds (of about 100 m in diameter) scattered on the flat land at fairly regular intervals (shown on the aerial photograph, Fig. 4), which are surrounded by a large ring of thick vegetation including tall trees. Such a discontinuous forest structure is quite comparable with the forest formations of the Sri Lanka dry zone, where "villus" and their large ring of tall forest are separated by flat areas of closed dry forest (see Hladik and Hladik, 1972). In the Morondava area, the distribution of some prosimians species, especially *Microcebus coquereli* (Petter *et al.,* 1971; see Chapter 4) is related to local changes in forest structure and composition.

Another type of major structural change is due to the proximity of the sea. Toward the west, the forest becomes gradually lower and more open, and apparently includes a smaller number of species. Nevertheless, in this area, the different nocturnal prosimians have been observed, although at low densities (Chapter 2).

The forest ends abruptly along a flat sandy and muddy open strip of land locally called "Sira Sira" (i.e., salty area). This sort of lagoon (Fig. 1B) is partly flooded by fresh water during the rainy season.

On the other side of the Sira Sira and bordering the sea is a mangrove formation including (as in other mangrove communities in the tropics) only

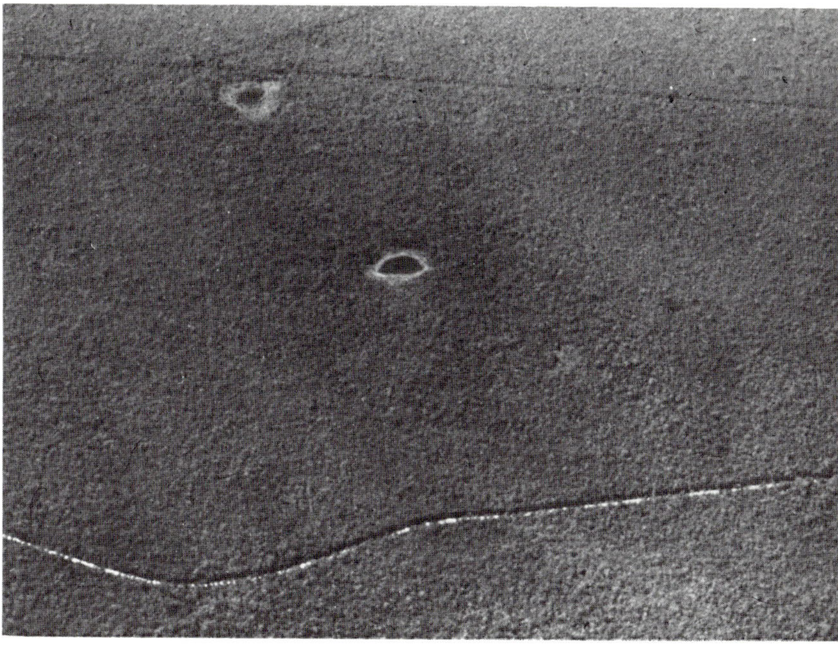

Fig. 4. An aerial view of the Marosalaza forest including some temporary ponds. In addition to the main road, old transects in the forest made for oil prospecting are visible and were used during the field study to census prosimian populations on a large scale (see Chapter 2).

a few tree species (Humbert, 1965). No prosimian has ever been observed there (P. Charles-Dominique and R. Albignac, personal communication). In Madagascar, diurnal prosimians have been observed in the mangroves of the East Coast (J.J. Petter, personal communication) which might be used only as part of their range. By contrast, on the West Coast of Africa, one primate population (*Cercopithecus aethiops*) which usually inhabits the dry forest, has colonized the mangrove by adapting its diet to crabs and a few plant species (Galat and Galat-Luong, 1977).

FLORISTIC COMPOSITION AND STRUCTURE OF THE FOREST IN THE STUDY AREA

The study area is located in the closed deciduous forest at about 3 km from the Sira Sira (Fig. 1B). This part of the forest has been protected during the last decades, and kept as a forest reserve adjacent to the sisal plantation. Most of the observations were carried on at about 0.5 km from the plantation, in an area of 30 hectares where footpaths were opened every 40 m (Fig. 5), with a minimum disturbance to the vegetation. A temporary pond is included in the south section of the study area.

Species Diversity

The floristic composition was analyzed by J.P. Abraham along two transects of 2200 m^2 (440 × 5 m) and 2700 m^2 (540 × 5 m), respectively. A total of 151 plant species were recorded on the 0.49 ha survey. An additional transect of 3400 m^2 (340 × 10 m) was made by C.M. Hladik, allowing comparisons of the densities of the most common trees and some records of liana distribution. All plant specimens collected during those botanical surveys, as well as the specimens related to prosimian feeding behavior, collected by the different investigators of the Morondava program, have been included in the collection of the "Service des Eaux et Forêts de Madagascar" (Tananarive) and in the Herbarium of the Muséum National d'Histoire Naturelle (Paris). About 200 plant species were collected. They are listed according to alphabetic order of families and genera in Appendix I. The list of corresponding vernacular names is given in Appendix II.

The plant species found in the first transect (2200 m^2) are listed in Table I, in decreasing order to their total basal area. The basal area of each tree is defined as the section area at 1.30 m above ground level; this measure provides a fairly accurate estimate of potential production. The number of individual trees in different girth and height classes is shown in this table. Plants below 30 cm girth (including tree saplings, lianas, shrubs, and herbs) have only been counted in the two first quadrats of the transect (i.e., 50 m^2) or are merely mentioned as a new species in the rest of the quadrat.

In Table II are listed all the liana species and only the most important trees not already included in Table I, after the data of J.P. Abraham's second transect and the record of all large trees potentially important as food for prosimians.

A comparison of our results on the total basal area of the trees in the study area with that of different plots surveyed by the Food Agricultural Organization (Anonymous, 1972; Table III; Fig. 1B) does not show any significant difference. Thus, in spite of the absence of some large specimens of Harofy, *Commiphora* spp. (24. . .)* and Hazomalany, *Hernandia voyroni* (87) which had been selectively exploited for timber in the 1960s, the study area is quite representative of the West Coast forest. The

* The numbers in parentheses refer to the checklist of species (Appendix I).

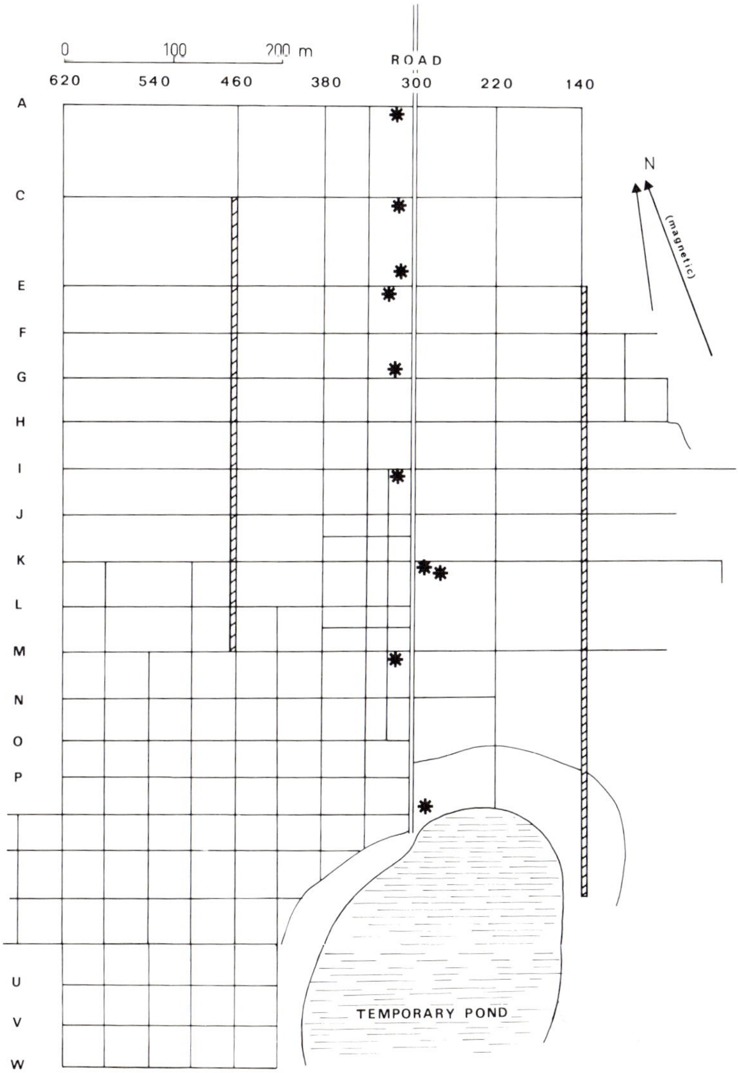

Fig. 5. The study area in the Marosalaza forest (see Fig. 1 for general location). Trails (solid lines) are referred to by numbers and letters. Most trails were opened every 40 m, parallel and perpendicular to the road. This preexisting narrow road leads from the sisal plantation (200 m north of trail A) to the temporary pond which is totally dry at the end of the dry season. The two transects of J.P. Abraham are shown by hatched areas. The transect surveyed by C.M. Hladik was located along the road between trails E and M. Stars indicate locations of the ten sampling baskets used for analysis of litter fall production.

TABLE I

Availability and Phenology of Plant Species in a Plot of Marosalaza Forest[a,b]

PLANT SPECIES	VERNACULAR NAME	NUMBER OF INDIVIDUALS ACCORDING TO GIRTH CLASSES (cm) 0	10	20	30	40	50	60	70	80	90	100 &+	TOTAL BASAL AREA (cm²)	INDIVIDUALS IN HEIGHT CLASSES (m) 1	5	&+	MONTHLY PHENOLOGY A S O N D J F M A M J J (1973 / 1974)
Securinega seyrigii	ANATSIKO	4	–	–	53	23	4						10000	3	8	73	. . . .XO. . . . .x. . . .
Adansonia fony	RENIALA	–	–	1	–	–	–	–	–	–	–	1	7000	–	–	2	.X. . . .O. . . .X.X - - - -
Baudouina fluggeiformis	MANJAKABETANY	1	–	1	4	4	3	–	2	2	1		4500	–	–	18	. . .X.X.X. . . -O- . . .
Dalbergia sp.	MANARY	7	–	–	4	4	3						1800	4	2	12	. . . .
Commiphora arafy	HAROFY	–	–	–	1	–	–	–	–	–	–	1	1250	–	–	2	. . .XO.O . . . .
Delonix boiviniana	FANGOKY	–	–	–	–	–	–	–	–	–	–	1	1000	–	–	1	. . . .X.X. . . . O. . . .
Dalbergia sp.	MANARIBERAVY	–	–	–	1	–	–	–	–	1			700	–	–	2	. . . . .O. . .
Dalbergia greveana	MANARIBAOMBY	–	–	–	–	4	2						1150	–	–	6	. .X.X. . .O.O. . . . . .
Strychnos decussata	HAZOMBY	20	–	7	7								700	18	3	13	. . . .X.O. . . . .
Cedrelopsis sp.	MANTAORA	7	–	–	2	2	1						800	3	4	5	. . . .
Operculicarya gummifera	SAKOAMBANDITSY	1	–	1	–	–	–	1					350	–	1	2	. .X. X.O. . . . .
Breonia sp.	HOLOMPANGA	–	–	–	–	–	–	1					350	–	–	1	. . . . .O. . . .
Diospyros humbertiana	KARIMBELALAHY	–	1	2		1							250	–	–	5	. . . . . .O.O. . . . . .
Grewia sp.	VALO	–	–	2		1							250	–	–	4	. . . . .X. . . . .
Turraea sp.	TSILAVONDRIA	1	–	–	–	–	1						250	–	1	1	. . . . .X. . . .X. . . .
Grewia cyclea	SELIBE	2	–	1	–	1							150	–	2	2	. . .X.X.X. . . . .O.O - - -
Pongamiopsis pervilleana	HAZOMBOATANGO	–	–	1	–	1							150	–	–	2	. . . .X. .O. . . . . . .
Berchemia discolor	TSIANDALA	2	–	–	–	–	1						250	2	1		. . .X.X.O. . . . .
Terminalia mantaliopsis	TALINALA	–	–	–	1								100	–	–	1	. . .x. .XO. O. . .X. .X. .

Broussonetia greveana	VORY	–	–	–	1		100	–	–	1
?	LALIANGALA	1	–	–	1		100	1	–	1
Ochnaceae	SAVONERA	–	–	–	1		100	–	–	1
Givotia madagascariensis	FARAFATSY	–	–	–	1		100	–	–	1
?	HAZOMENALALA	28	–	–	2		200	9	21	
Bauhinia sp.	BANAKY	31	–	–	1	Under 30 cm girth ↓	100	22	10	
Buxus madagascariensis	TSIVOANINOM-BEZO	2	2					1	–	3
Phyllanthus cf. decaryanus	MANDIFIKA	52						18	34	
Croton aff. jennyanum	VOLAFOTY	8	–	2				1	7	2
Croton menebeensis	SOAHY	?						?		
Chloroxyllon falcatum	MANDAKOLAHY	–	–	1		(Large tree in other plots)		–	–	1
Cedrelopsis sp.	MAMPANDRA	1	–	1				1	–	1
Apocynaceae	HAZOMIPIKA	1	–	1				1	–	1
Rubiaceae		–	–	1				–	–	1
Grewia triflora	SELY	–	1			(Large tree in other plots)		–	–	1
Zanthoxyllum tsihanimposa	MONONGO	–	1			(Large tree in other plots)		–	–	1
Noronhia boinensis	TSILAITSY	–	1					–	–	1
Colubrina alluaudii		–	1					–	–	1
Colvillea racemosa	SARONGAZA	–	1			(Large tree in other plots)		–	–	1
Tiliaceae	SELIMAVO	–	1					–	–	1
?	TSIENGENA	–	1					–	–	1
Apaloxylon tuberosum	TALAMENA	–	1					–	–	1
Neobeguea mahafaliensis	HANDY	1				(Large tree in other plots)		–	–	1
Grewia voloina	SELIMALITSY	2						–	1	1

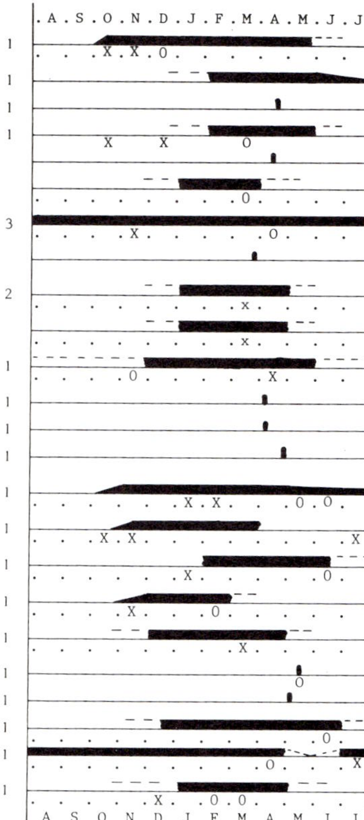

(*Continued*)

TABLE I *(Continued)*

PLANT SPECIES	VERNACULAR NAME	NUMBER OF INDIVIDUALS ACCORDING TO GIRTH CLASSES (cm) 0	10	20	30 40 50 60 70 80 90 100 &+	TOTAL BASAL AREA (cm²)	INDIVIDUALS IN HEIGHT CLASSES (m) 1	5	&+	MONTHLY PHENOLOGY A S O N D J F M A M J J (1973 / 1974)
Capuronia madagascariensis	KITATAMALANDY	1					–	–	1	X . X . 0 . 0
Diporidium pervilleanum	BORAMENA	4					–	–	4	X . 0
Hibiscus cf. diplocrater	MAINATY	1	1				–	1	1	0
Nesogordonia cf. ambalalensis	REFEKO	1					–	–	1	
?	PITIKALA	1					–	–	1	
Celastraceae	TSIVOANIZAO	1					–	–	1	
?	HAZOFIA	1					–	–	1	
Diospyros sp.	HAZOBOENGA	3	–	1			3			0
Grewia sp.	SELIMBIVY	3					–	3		X 0
Grewia sp.	SELIFOTIVORY	4					1	3		
?	SOMPATSALA	5					2	3		
Pyrostria sp.	FATIKAHITRA	2					–	2		0
Hildegardia erythrosyphon	VONOA	1			(Large tree in other plots)		–	1		X . X . 0 . 0
Phyllanthus rhomboidalis	MANANDRA	1					–	1		0
cf. Deuteromallotus	MAMIHY	1					–	1		X
Memecylon sp.	TAHILA	2					1	1		
Rubiaceae	TSINJOVY	1					–	1		
Euphorbia antso	ANTSO	1					–	1		
?	TSIVOANINO	1					–	1		
?	(A.74)	1					–	1		
Hymenodictyon cf. louhavate	BEHOLITSE	1			(Large tree in other plots)		1			X . 0 – – – 0
Psorospermum sp.	HARONGAMPANIHY	1					1			X
Phyllarthron bernierianum	TOHIRAVY	1					1			
Diospyros sp.	MANDRAVASARETRA	1					1			
Terminalia sp.	AMANIOMBY	1					1			
Ebenaceae	PIRIPITSOKALA	1					1			X
Acanthaceae	VELATRA	1					1			
										A S O N D J F M A M J J

LIANA SPECIES										A S O N D J F M A M J J
Cyphostemma pachypus		-	-	-	1		-	-	1	. . . . X . O . . . .
Elachyptera minimiflora	HAZONTA	4					2	-	2	. . . . X . . X . . . . . .
Asclepiadaceae	(A.24)	5					-	-	5	. . . . O . . . .
Leguminosae	VAHINTA	5					-	-	5	
Acanthaceae	VOLILY	12					3	9		
Cynanchum sp.	VOAHAVORY	1					-	-	1	
Vanilla cf. montagnacii	VAHIMALO	1					-	-	1	
Byttneria oligantha	VAHINAMALONA	1					-	1		. . . . X . X . . . . . O . .
?	(A.62)	1					-	1		
?	(A.73	1					-	1		
?	(A.77)	1					-	1		

[a] The phenological data are summarized according to a review of specimens of the Paris Herbarium and illustrated as follow :

x = flower buds X = flowers 0 = ripe fruits

leaves for evergreen species; new leaves – leaf fall: leaves for deciduous species

partial phenological data; leaves present at the collecting time during this study

[b] Size classes and phenology of the plant species recorded by J. P. Abraham in the first transect (440 m × 5 m = 2200 m²). Trees are initially listed according to increasing order of girth and height classes. Total basal area is not listed for those species with a girth under 30 cm. Species with a girth exceeding 30 cm account for most of the biomass. Liana species are listed separately.

TABLE II

Phenology of Plant Species (Complementary List)[a]

TREE SPECIES	VERNACULAR NAME	A S O N D J F M A M J J
Cussonia sp.	VOANTSILANA	
Kigelianthe sp.	TSOMONTSOY	
Adansonia grandidieri	RENIALA	. . 0 . 0 . .0? . . . . X . .
Commiphora stellulata	HAROFIMENA	. . . . X . 0 . 0 . . . .
Commiphora cf. pterocarpa	HAROFIBOY	. . . . X? . . . . 0 . . .
Commiphora aff. mafaidoha	HAROFIFOTSY	. . . . 0 . . . .
Cordyla madagascariensis	ANAKARAKY	. . . X . . . . 0 . .
Cassia sp.	SARONGAZALA	. . . . 0 . . . . .
Delonix floribunda	MALAMASEFOY	. . . X . X . X . . . 0 . . . . .
Tamarindus indica	KILY	. . . . X . . . . 0 . . .
Tetraptaerocaryon geayi	VAOVY	. . . . 0 . . .
Boscia madagascariensis	SOMANGIBE	. . . . . X? . . . . . X? . . .
Evonymus pleurostyloides	MARONONO	. . . . X . 0 . . . . . .
Terminalia tricristata	TALY	. . . . . . X . . . 0 . 0 . 0 . 0 . .
Terminalia bovinii	KOFAFA	. . . X . X . . 0 . 0 . 0 . . . .
Securinega sp.	HAZOMENA	. . . .
Hernandia voyronii	HAZOMALANY	. 0 . . . . X . . . . .
Fœtida asymetrica	MENAMBAHY	. . . . . . X . . . . 0 . . . .
Albizia sp.	HALOMBORO	. . . . X . X . . . 0 . . . . .
Chadsia flammea	REMOTY	. . . X . . . . 0 . .
Chadsia versicolor		. . . . X . . . . 0 . .
Colubrina decipiens	TRATRAMBORONDREO	. . . . . . X . X . . . . 0 . 0 . .
Rubiaceae	PAPAHOLAHY	. . . . 0 . . .
Rubiaceae	TSIHATAIMBALY	. . . .
Rubiaceae	VOAFOTAKA	. . . . 0 . . . . .
Nesogordonia cf. micrantha	HAZOMENARANO	. . . .
Majidea zangabarica		. . . . XO . . .
Capurodendron rubrocostata	NATOKORONONO	. 0 . X . . . . 0 . . . .

LIANA SPECIES	VERNACULAR NAME	. A . S . O . N . D . J . F . M . A . M . J . J .
Asparagus vaginellatus	TSIMATY	. . . . X . . . . . . 0 . 0 . . .
Cryptostegia madagascariensis		

TABLE II (*Continued*)

LIANA SPECIES	VERNACULAR NAME
Combretum coccineum	
Metaporana sericosepala	KILILO
Dioscorea sp.	BABO
Anisocyclea grandidieri	TSIDIKANDAMBOLAHY
Desmonema uviforme	BOKATSIFAKA
Gouania lineata	VAHINDA
Landolphia cf. obliquinervia	KABOKA
Asclepiadaceae	VAHIRONTO
Asclepiadaceae	(A.133)
Rubiaceae	VAHIMANTSINA
?	BETONDRO
?	(A.101)
?	(A.131)
cf. Tragia triverneana	
Dalechampia cf. clematidifolia	VAHIMBARAONGO
Dalechampia cf. subternata	MIATELORAVY
cf. Acalypha perrieri	

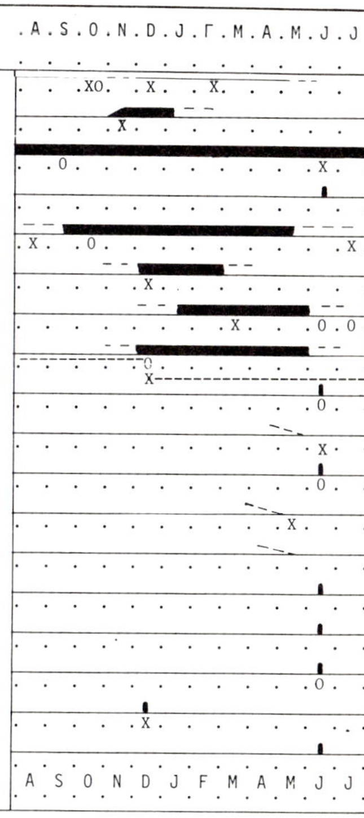

[a] Complementary list of plant species according to the second transect (J. P. Abraham) and other specimen collections in the study area. Data on phenology are illustrated as in Table I.

first transect (2200 m^2, 80 species including 12 liana species) and the second transect (2700 m^2, 102 species including 10 liana species), completed by other collections in the study area, are presently the most complete species analysis available for this type of plant formation in Madagascar.

For some common species, the total basal area measured in different transects (Table IV), shows a gradual change across the study area. Such a lack of homogeneity characterizes any tropical forest. These local variations are very important in term of distribution of the food available in the home ranges of different individuals of a given species and, in some cases, different food species may play the same role for different animals in nearby places (see Hladik, 1977; Rudran, 1978).

In the Marosalaza forest, species diversity is slightly higher than in the deciduous forest of Wilpattu (Sri Lanka) (Hladik and Hladik, 1972) which is a very comparable habitat. There are 26 species over 30 cm girth in the

TABLE III

Total Basal Area (m^2/ha) of All Plant Species over 20 cm Girth in Different Locations (Fig. 1) of the Forest of the West Coast of Madagascar

Forest locations	Sample size (ha)	All stems over 20 cm girth	30 cm girth	60 cm girth
Marofandilla forest, F.A.O. survey 21	4	15.6	—	7.0
Marofandilla forest, F.A.O. survey 22	4	19.6	—	10.3
South of Tsiribihina river, F.A.O. survey 23	4	22.5	—	11.9
North of Tsiribihina river, F.A.O. survey 24	4	17.9	—	9.3
Marosalaza forest, Hladik's survey	0.34	18.5	—	13.8
Marosalaza forest, Abraham's survey	0.22	—	13.0	6.2

transect of 2200 m^2 of Marosalaza, while in Wilpattu, only 19 species were recorded in a transect of similar shape and size. In both forests, the most common species (with about 250 individuals above 30 cm girth per hectare) is a small tree, with persistent leaves, of the Euphorbiaceae family: the Anatsiko, *Securinega seyrigii* (79) of Madagascar, and the Veera, *Drypetes sepiaria* of Sri Lanka. Other tree species of these two dry forests located on opposite sides of the Indian Ocean and used as food by primates also have ecological homologs. According to our observations, the tamarind tree or Kily, *Tamarindus indica* (40) and the Malagasy plum tree or Lamoty, *Flacourtia indica* (83), together with the tuberous plant Tavolo, *Tacca leontopetaloides* (165), are the only species present in both forests.

TABLE IV

Total Basal Area (m^2/ha) and Local Variation of Some Common Tree Species in the Marosalaza Forest

Tree species	West[a] (2200 m^2)	Central[b] (3400 m^2)	East[a] (2700 m^2)
Anatsiko, *Securinega seyrigii* (79)	4.55	1.94	1.74
Reniala, *Adansonia* spp. (20. . .)	3.27	8.56	—
Manjakabetany, *Baudouina fluggeiformis* (31)	3.25	1.50	0.01
Hazomby, *Strychnos decussata* (93)	0.50	0.62	0.26
Harofimena, *Commiphora stellulata* (25)	0.59	1.00	1.11

[a] Survey by J.P. Abraham.
[b] Survey by C.M. Hladik.

PHENOLOGY AND FOOD AVAILABILITY

The phenological patterns of the most common plants of the study area were analyzed according to the measures of leaf fall (Fig. 8). The leaves of 27 species were obtained in sufficient quantity for analysis variations, from a sample of 8 litter traps. The leaves of these species accounted for 95% of the total leaf fall. Due to the arbitrary location of the litter traps and the small collecting area (4 m^2), it is not possible to compare the absolute quantities of leaves, but only the different patterns of leaf fall. These patterns can be classified in four groups as follows:

A. *Evergreen species:* The small number of species with xerophytic leaves present in dry season and falling in small quantity throughout the year (Fig. 8A) includes some common bushes and small trees such as the Anatsiko, *Securinega seyregii* (79), the Iron Wood or Hazomby, *Strychnos decussata* (93), the Tsivoaninombezo, *Buxus madagascariensis* (29), and the Hazoboenga, *Diospyros* sp. (59). The vine Hazonta, *Elachyptera minimiflora* (88), is of particular interest since it is the main (if not unique) food species of an homopteran larvae producing a sweet secretion of importance in the diet of *Microcebus coquereli* (Chapters 2 and 4). The flowering and fruiting periods are very variable among the different species of this group (Table I).

B. *Early deciduous species:* This group includes the most common large trees such as the Baobabs or Reniala, *Adansonia* spp. (20. . .), the Harofy, *Commiphora* spp. (24. . .), and some Sterculiaceae and Rubiaceae. All these trees are totally leafless as early as May (Fig. 8B). Most species have a flowering period preceding the new leaf flush, the Baobabs being the most precocious. The different early deciduous species produce large flowers rich in nectar and/or fleshy parts just before, or at the beginning of the rainy season and thus provide an important amount of food immediately available to animals. The fruits of these ripen a short time after the leafing period (Table I).

C. *Slow deciduous species:* Intermediate patterns of progressive leaf shedding (Fig. 8C) occur in at least three common trees: the Selibe, *Grewia cyclea* (166), the Sely, *Grewia triflora* (169), and the Tititamalandy, *Capuronia madagascariensis* (94). Among them, only the Selibe flowers before leafing: this large common Tiliaceae, the aspect and ecology of which are very similar to those of *Adina cordifolia* (a Rubiaceae of the dry forest of Sri Lanka), also produces a dry inedible fruit in June. Both species are cropped by primates mainly for their flowers.

D. *Late deciduous species:* Most of these species were still shedding leaves in July (Fig. 8D); accordingly, they are bare for only a short period and thus somewhat similar to those of group A. This group includes more than one-third of the recorded species which all flower when new leaves are present. Among them, the Talinala, *Terminalia mantaliopsis* (48), is important as a permanent supply of gums as food for *Phaner furcifer* (see Chapter 3).

The overall phenological data presented in Tables I and II have been based on patterns observed in 1973 and 1974. Complementary information was available in the Herbarium of the Muséum National d'Histoire Naturelle (Paris) after a review of all the herbarium samples previously collected in the Morondava area. In these cases, data from different years have been cumulated. Due to the seasonal changes from year to year, which might be important in spite of a regular climatic pattern, the yearly variation in plant phenology must be considered. For example, in the nearby forest of Marofandilla, three plant species have been observed during 5 years (Gachet, 1971). Different individuals of a given species definitively had the same phenological pattern, but occasional variations in the specific pattern may be attributed to variations in the distribution of the rainfall. Some of these changes may be important, such as the absence of flowers following an extremely dry year.

Accordingly, the information grouped in Tables I and II might be considered as a general background of the phenological events in Marosalaza forest, allowing an overall appreciation of the influence of primary production on the life cycles of the animals presented in other chapters of the present volume.

ALKALOID SCREENING IN LEAVES

Records on plant species diversity, density, and distribution, as well as observations on plant phenology (leafing, flowering, and fruiting) provide a general analysis of food available to animals as a function of space and time. However, what plants and which plant parts are edible, and to what species of prosimians? Since alkaloids have been shown to be responsible for plant toxicity, tests have been made on the dry leaf samples from Marosalaza forest (for methods and discussion, see Hladik and Hladik, 1977).

Results are presented in Table VI. The positive results (+ + + and + +) show a relatively low percentage, 15% (19 positive out of 127 species) similar to that of other closed tropical forests. Most positive results are found in families generally well known for the occurrence of alkaloids: Apocynaceae, Euphorbiaceae, Menispermaceae, and Solanaceae. How-

TABLE VI

Results of Alkaloid Tests with Mayer's (M) and Dragendorff's (D) Reagents on Dry Specimens of Leaves [a]

		M	D
Acanthaceae	1	−	−
	2	−	±
	5	−	±
Amarantaceae	6	±	±
Anacardiaceae	7	−	−
Apocynaceae	8	−	−
Araliaceae	10	−	−
Asclepiadaceae	11	−	±
	12	±	±
	13	−	−
	14	±	±
Bignoniaceae	17	−	−
	18	−	−
	19	−	−
Bombacaceae	20	−	−
Burseraceae	24	−	++
	25	−	−
	27	−	−
	28	−	−
Buxaceae	29	+++	+++
Caesalpiniaceae	30	±	±
	31	±	±
	35	−	−
	36	−	−
	37	−	−
	38	−	−
	40	−	−
	41	−	++
Capparidaceae	42	−	−
Celastraceae	43	−	±
	44	−	−
	45	−	−
Combretaceae	46	−	++
	48	−	+
	49	±	++
Convolvulaceae	52	−	−
Cyperaceae	53	−	−
Dioscoreaceae	54	−	−
Ebenaceae	55	−	−
	56	−	−
	57	−	+
	59	±	++
Euphorbiaceae	62	−	−
	63	−	−
	64	−	±
	65	−	−
	66	−	−
	67	+++	+++
	70	±	++
	71	±	++
	72	−	−
	73	++	++
	74	−	−
	75	−	−
	76	±	++
	77	±	++
	78	±	++
	79	+	+
	80	+	++
	81	−	−
	82	−	−
Flacourtiaceae	83	−	−
	84	−	−
Graminae	86	+	+
Hernandiaceae	87	−	−
Hippocrateaceae	88	±	+
Hypericaceae	89	±	+
Lecythidaceae	90	±	±
Liliaceae	91	−	−
Loganiaceae	92	−	−
	93	+	++
Lythraceae	94	−	++

(*Continued*)

TABLE VI (*Continued*)

	M	D
Malvaceae		
95	–	–
97	–	+
98	–	±
Melastomaceae		
100	–	±
Meliaceae		
103	–	+
104	+	+
105	–	–
106	±	±
Menispermaceae		
107	+++	+++
108	±	±
Mimosaceae		
109	–	±
110	–	–
111	–	±
114	–	++
Moraceae		
115	–	–
Ochnaceae		
116	–	–
117	–	–
Oleaceae		
119	–	–
Papilionaceae		
124	–	–
125	–	–
127	–	–
128	±	±
129	±	±
Pedaliaceae		
132	–	+
Rhamnaceae		
133	+	+
134	+	+
135	–	–
136	+	+
Rubiaceae		
137	–	–
138	–	–
139	±	±
140	±	±
141	–	–
142	–	–
Rutaceae		
154	±	±
Sapindaceae		
155	±	±
156	+	+
157	+	++
Sapotaceae		
158	–	–
Solanaceae		
159	++	++
Sterculiaceae		
160	–	–
161	–	–
162	–	–
163	–	–
164	±	±
Taccaceae		
165	±	±
Tiliaceae		
166	–	–
167	±	±
168	–	–
169	±	±
170	–	–
173	–	–
Turneraceae		
177	–	–
Verbenaceae		
178	+	+
179	±	±
Vitaceae		
180	–	–
181	–	–

[a] + + + = important precipitate; + + = strongly positive; + = slightly positive; ± = traces; – = no reaction. Plant species (referred to as species number in Appendix I) are listed in the alphabetic order of families.

ever, in Marosalaza forest where most of the species are deciduous, it appears that the evergreen species show positive results, for example, the small tree Tsivoaninombezo, *Buxus madagascariensis* (29), and a liana Tsidikandambolahy, *Anisocyclea grandidieri* (107). In terms of plant strategy, a high alkaloid content might be considered as a deterrent against predators (Janzen, 1969).

In fact, the degree of alkaloid toxicity of the various plant species to prosimians is still unknown. Considering the overall results, the situation in the Marosalaza forest appears similar to that of other closed tropical forests: leaf-eating primates feed on a set of plant species in which the proportion of plants containing alkaloids is equivalent to that of the forest as a whole (C.M. Hladik, 1978). Thus, in these cases, alkaloids do not appear to affect food choices.

The major effect of plant alkaloids seems to apply to invertebrate populations which are the main consumers of leaves (Leigh and Smythe, 1978). This would explain why the evergreen species of the Marosalaza forest which are more exposed to predators during the dry season present a higher alkaloid content. Similarly, in open areas where a limited number of plant species may form large targets for insects, a higher percentage of alkaloids is generally found. This was the case in the survey of Debray *et al.* (1971) in the open area of the forest of the south west of Madagascar (including many herbaceous species) where 22% of the plants yielded positive results (21 positive leaves out of 94 tested species).

DISCUSSION

In the Marosalaza forest the overall production, in terms of litter fall, is one of the lowest among tropical forests. Plant diversity is limited as compared to that of rain forests; however, about 200 plant species have been collected in the study area. In spite of the important seasonality, and possibly poor sandy soils, plant species diversity provides a large variety of food available for prosimians because leafing, flowering, and fruiting times differ according to the species.

These phenological patterns dependent on abiotic factors such as the long period of drought, might also be modulated by various biotic factors, especially interactions with animal species in terms of pollination, seed dispersal, and consumption.

In this respect, the leafing patterns classified in the Marosalaza forest from evergreen to various deciduous categories, might be either a response to drought or the result of selective pressure by folivorous animals. Leaves of evergreen species, which may additionally develop morphologic adaptations (spiny, waxy, or tough leaves) show an increased amount of toxic or unpalatable substances (alkaloids and/or other secondary compounds). The potential leaf consumers might be essentially the invertebrate fauna, which rapidly increases in biomass after most deciduous trees grow leaves (see Chapter 2). In contrast, the vertebrate fauna does not seem to be a threat for plant species since only a small proportion of leaves are eaten (see discussion about folivores in C.M. Hladik, 1978).

Flowering and fruiting patterns are more subtle than leafing patterns. Flowering times which in many plants follow the long dry season, might be also influenced by various pollinating agents and their times of activity (as discussed by Janzen, 1967, and more recently by Frankie *et al.,* 1974). Besides insects, prosimians may act as pollinators particularly in plants which have a high content of pollen and/or nectar, for instance, the Boy, *Delonix floribunda* (38), and the Maronono, *Evonymus pleurostyloides* (43), consumed respectively by *Cheirogaleus medius* and *Microcebus murinus* (Chapter 2). These plants flower at the beginning of the rainy season when these two prosimian species resume activity after the period of lethargy. Other cases of such flowers pollinated by vertebrates have been observed in southern Africa (*Protea* spp. by rodents, Wiens and Rourke, 1978), and especially in Madagascar (*Ceiba pentandra* by *Lemur mongoz,* Sussman, 1978). The trophic relations between animals and plants depend on plant morphology and biochemical composition, as well as on biological cycles (Baker and Baker, 1975; Baker, 1978).

Fruiting patterns are more variable and difficult to classify than flowering ones because, for some plant species, ripe fruits immediately follow flowering, while in others, fruits require a long time to mature (sometimes as much as 1 year). As in the case of flowers, biochemical composition and palatability of fruits may result from interactions with vertebrates. At the beginning of the rainy season, the fruits of several plant species in the Marosalaza forest have a high protein content (compared to the mean value of most tropical fruits) and are particularly attractive to the frugivore/insectivore prosimian species during the period of relative insect scarcity (Chapter 2). As a result, any plant with a relatively high protein content in fruits will benefit from seed dispersal in the feces of prosimians. [For general discussion on seed dispersal, see Ridley (1930); concerning primates, see C.M. Hladik and A. Hladik (1967) and A. Hladik and C.M. Hladik (1969).] In this way, coevolution between seed dispersal agents and plants producing edible fruits might determine differences in biochemical composition and time of ripening between closely related plant species (Hladik and Hallé, 1979).

In the Marosalaza forest, the long period of drought is obviously the major factor affecting the biological cycles of both plants and animals. Nevertheless, as in other tropical forests, an understanding of the interrelations between the many plant species and their predators, pollinators or seed dispersors, as a result of a long history of coevolution, allows a better comprehension of biology of prosimian species in relation to the contrasting conditions of this environment.

APPENDIX I

List of the Plant Species of Marosalaza Forest (North of Morondava) with Vernacular Names, Life Forms, Herbarium Specimen Numbers with References to Collectors,[a] and Phenological Notes

ACANTHACEAE

1. *Hypoëstes* sp., HAZONDOLO, Shrub, A.83, flowers in May
2. *Hypoëstes* sp., VELATRA, Shrub, A.80, flowers in May
3. *cf. Justicia* sp., VELATRATAKA, Herb, A.132, flowers in June
4. *cf. Oplonia* sp., Tree, CD.1, flowers in November
5. VOLILY, Shrub, A.8, sterile in April

AMARANTACEAE

6. *Achyranthes aspera* L., VAOTOMPAOSA, Herb, A.143, fruits in June

ANACARDIACEAE

7. *Operculicarya gummifera* (Sprague) Cap., SAKOAMBANDITSY, Tree, H.3, fruits in November; H. 23, fruits in December; A.39, sterile in May

APOCYNACEAE

8. *Landolphia cf. obliquinervia* Pichon, KABOKA, Liana, H.33, fruits in December
9. HAZOMIPIKA, Tree (?), A.23, sterile in May

ARALIACEAE

10. *Cussonia* sp., VOANTSILANA, Tree, A.156, sterile in May

ASCLEPIADACEAE

11. *Cryptostegia madagascariensis* Boj., LOMBIRY, Liana, H.17, flowers in December
12. *Cynanchum* sp., VOAHAVORY, Liana, A.41bis, flowers and fruits in April
13. KOMPOTSOY, Liana, H.39, flowers in December
14. VAHIRONTO, Liana, A.138, fruits in June
15. Liana, A.24, fruits in April
16. Liana, A.133, flowers in June

(*Continued*)

BIGNONIACEAE

17. *Kigelianthe* sp., TSOMONTSOY, Tree, A.86, sterile in June
18. *Phyllarthron bernierianum* Seeman, TOHIRAVY, Tree, A.9, sterile in April
19. *Stereospermum euphorioides* Perr., MANGARAHARA, Tree, A.129, flowers and fruits in December

BOMBACACEAE

20. *Adansonia fony* Baill., RENIALA, Tree, A.48, flowers in April
21. *Adansonia grandidieri* Baill., RENIALA, Tree, H.27, flowers in December
22. *Adansonia za* Baill., Tree likely to be present (photo collection H. 849 N).

BORAGINACEAE

23. *Ehretia* sp., SOFIKOMBA, Tree, A.104, sterile in May

BURSERACEAE

24. *Commiphora arafy* Perr., HAROFY, Tree, H.11, fruits in December; A.46, sterile in May
25. *Commiphora stellulata* Perr., HAROFIMENA, Tree, A.108, sterile in May
26. *Commiphora cf. pterocarpa* Perr., HAROFIBOY, Tree, A.161, fruits in May
27. *Commiphora aff. mafaidoha* Perr., HAROFIFOTSY, Tree, A.100, sterile in May
28. *Commiphora simplicifolia* Perr., SEKATSY, Shrub, A.145, sterile in May

BUXACEAE

29. *Buxus madagascariensis* Baill., var. *tropophylla,* TSIVOANINOMBEZO, Tree, JJP.20, flowers in November; A.2, fruits in March

CAESALPINIACEAE

30. *Apaloxylon tuberosum* Vig., TALAMENA, Tree, A.72, fruits in June
31. *Baudouina fluggeiformis* Baill., MANJAKABETANY, Tree, H.15 and H.18, flowers in December; A.42, fruits in April
32. *Bauhinia aff. grandidieri* Baill., Tree, A.116, sterile in June
33. *Bauhinia* sp., BANAKY, Tree, A.16, fruits in March
34. *Cassia* sp., SARONGAZALA, Tree, A.159, fruits in March
35. *Colvillea racemosa* Boj., SARONGAZA, Tree, H.59, sterile in January; A.55, flowers in March
36. *Cordyla madagascariensis* Vig., ANAKARAKY, Tree, A.119, fruits in June
37. *Delonix boiviniana* (Baill.) Cap., FANGOKY, Tree, JJP.2, flowers in November; H.16, flowers in December; A.44, fruits in April
38. *Delonix floribunda* (Baill.) Cap., Tree, BOY, H.14, flowers in December; MALAMASEFOY, A.112, sterile in June
39. *Sakoanala menabeensis* Peltier, MATAMBELO, Tree, A.152, flowers in May
40. *Tamarindus indica* L., Tree, KILY, H.1, flowers in November
41. *Tetrapterocarpon geayi* Humb., Tree, VAOVY, A.162., fruits in June

CAPPARIDACEAE

42. *Boscia madagascariensis* (D.C.) Hadj.-Moust., SOMANGIBE, Tree, A.118, sterile in June

CELASTRACEAE

43. *Evonymus pleurostyloides* Loesn., MARONONO, Tree, H.61, flowers in January; GP.1, fruits in February
44. TSIVOANIZAO, Tree, A.64, sterile in June
45. REAMPY, Tree, A.110, sterile in May

COMBRETACEAE

46. *Combretum coccineum* Lam., Liana, JJP.15, flowers in November
47. *Terminalia bovinii* Tul., KOFAFA, Tree, JJP.28, fruits in March; A.102, sterile in June
48. *Terminalia mantaliopsis* Perr., TALINALA, Tree, H.28, fruits in December; H.29, flowers in December; A.54, sterile in May; TALISARA, A.166, sterile in June
49. *Terminalia tricristata* (Perr.) Cap., TALY, Tree, H.58, unripe fruits in January; A.150, fruits in May
50. *Terminalia* sp., AMANIHOMBY, Shrub, A.35, sterile in June

COMPOSITAE

51. *Elephantopus* sp., TAMBAKONJIRIKY, Herb, A.128, fruits in November

CONVOLVULACEAE

52. *Metaporana sericosepala* Verdc., KILILO, Liana, A.99, flowers in June

CYPERACEAE

53. *Cyperus obtusiflorus* Vahl, MITA, H.8, fruits in November

DIOSCOREACEAE

54. *Dioscorea* sp. BABO, Liana, A.130, sterile in June

EBENACEAE

55. *Diospyros aculeata* Perr., RELEFO, Tree, H.4, young fruits in November
56. *Diospyros aff. greveana* (Pierre) Perr., Tree, H.13, young fruits in December
57. *Diospyros humbertiana* Perr., KARIMBELALAHY, Tree, A.45, sterile in April
58. *Diospyros* sp., MANDRAVASARETRA, Tree, A.26, sterile in April
59. *Diospyros* sp., HAZOBOENGA, Tree, A.18, fruits in March
60. *Diospyros* sp., PIRIPITSOKALA, Shrub, A.36, sterile in April

ERYTHROXYLACEAE

61. *Erythroxylum* sp., MENAHIHY, Shrub, A.113, sterile in June

(*Continued*)

EUPHORBIACEAE

62. *cf. Acalypha perrieri* Leandri, Shrub, A.142, sterile in June
63. *Antidesma petiolare* Tul., Tree, JJP.17, fruits in November
64. *Bridelia pervilleana* Baill., KITATA, Tree, A.122, fruits in May
65. *Croton menabeensis* Leandri, SOAHY, Tree, A.4, flower buds in April
66. *Croton cf. jennyanum* A. Gris, VOLAFOTY, Tree, A.28, flower buds in April
67. *Croton cf. nobilis* Baill., ANDRIAMBOLAFOTY, Tree, JJP.29, flower buds in March; A.163, flower buds in May
68. *Croton* sp., Tree, JJP.26, flower buds in March
69. *Croton* sp., Tree, JJP. 9, flowers in November
70. *Dalechampia cf. clematidifolia* Baill., VAHIMBARAONGO, Liana, A.82, fruits in June
71. *Dalechampia cf. subternata* Muell.—Arg., MIATELORAVY, Liana, H.41, flowers in December
72. *cf. Deuteromallotus* sp., MAMIHY, Shrub, A.60, flower buds in May
73. *Euphorbia antso* M. Den., ANTSO, A.45bis, sterile in May
74. *Euphorbia cf. millii* Desmoulins, A.87, sterile in June
75. *Givotia madagascariensis* Baill., FARAFATSY, Tree, A.52, sterile in April
76. *Phyllanthus cf. decaryanus* Leandri, MANDIFIKA, Shrub, A.5, sterile in March
77. *Phyllanthus casticum* Willem, SANIRA, A.135, sterile in June
78. *Phyllanthus rhomboidalis* Muell.—Arg., MANANDRA, JJP.13, sterile in November; A.25, fruits in April
79. *Securinega seyrigii* Leandri, ANATSIKO, Tree, H.62, sterile in January; A.1, flower buds in March
80. *Securinega* sp., HAZOMENA, Tree, A.117, sterile in June
81. *Suregada cf. capuronii* Leandri, TSILAVONDRIA, Tree, A.31, flower buds in May
82. *cf. Tragia triverneana* Leandri, Liana, A.141, sterile in June

FLACOURTIACEAE

83. *Flacourtia indica* (Burm. f.) Merr., LAMOTY, Tree, JJP.21, fruits in November
84. *Homalium albiflorum* (Boiv. ex Tul.) Hoffm., HAZOMBO, Tree, H.44, flowers in December; A.85, flowers in May
85. *Physena sessiliflora* Tul., Shrub, H.53, fruits in January

GRAMINAE

86. *Pennisetum polystachyon* Sch., TSIBOSIBOSY, A.127, fruits in June

HERNANDIACEAE

87. *Hernandia voyroni* Jumelle, HAZOMALANY, Tree, A.164, sterile in May

HIPPOCRATEACEAE

88. *Elachyptera minimiflora* (Perr.) N. Hallé, HAZONTA, Liana, H.55, flowers in January; A.21, sterile in April

HYPERICACEAE

89. *Psorospermum* sp., HARONGAMPANIHY, Shrub, A.75, flowers in June

LECYTHIDACEAE

90. *Foetida asymetrica* Perr., MENAMBAHY, Tree, A.109, fruits in May

LILIACEAE

91. *Asparagus vaginellatus* Boj., TSIMATY, Liana, A.107, fruits in May

LOGANIACEAE

92. *Mostuea brunonis* Didr., SOFITAMBA, Shrub, JJP.1, flowers in November; A.98, sterile in June
93. *Strychnos decussata* (Pappe) Gilg, HAZOMBY, Tree, H.5 and H.54, flowers in November and January; H.25, fruits in December; A.17, sterile in April

LYTHRACEAE

94. *Capuronia madagascariensis* A. Lourteig, KITATAMALANDY, Shrub, A.53bis, sterile in May

MALVACEAE

95. *Gossypium brevilanatum* (Hochr.) Hutch., HASIKITOKY, A.165, flowers and fruit in May
96. *Hibiscus cf. diplocrater* Hochr., MAINATY, A.33, fruits in April
97. *Hibiscus macrogonus* Baill., HALAMPO, A.157, flowers in May
98. *Hibiscus vitifolius* L., A.147, fruits in June
99. *Perrierophytum* sp., A.136, fruits in June

MELASTOMACEAE

100. *Memecylon* sp., TAHILA, Shrub, A.14, sterile in April

MELIACEAE

101. *Cedrelopsis* sp., MANTAORA, Tree, A.13, sterile in April
102. *Cedrelopsis* sp., MAMPANDRY, Tree, A.30, sterile in April
103. *Chloroxylon falcatum* Cap., MANDAKOLAHY, Tree, A.40, flowers in April
104. *Neobeguea mahafaliensis* Leroy, HANDY, Tree, H.26, fruits in December; A.61, fruits in April
105. *Turraea* sp., Shrub, H.37, flowers in December
106. *Turraea* sp., HAZOMPODY, Shrub, H.42, flowers in December

MENISPERMACEAE

107. *Anisocyclea grandidieri* Baill., TSIDIKANDAMBOLAHY, Liana, A.97, sterile in June
108. *Desmonema uviforme* (Baill.) Diels, BOKATSIFAKA, Liana, H.20, flowers in December

MIMOSACEAE

109. *Acacia aff. drakei* Vig., FATIKY, Tree, A.66, sterile in June
110. *Acacia pervillei* Benth., Liana, H.48 and H.48bis, flowers in November
111. *Acacia sakalava* Drake, ROY, Tree, JJP.18, flowers in November

(*Continued*)

112. *Albizia* sp., HALOMBORO, Tree, A.160, fruits in April
113. *Dichrostachys arborescens s.l.,* Shrub, JJP.12ter, flowers in November
114. *Mimosa aff. menabeensis* Vig., Tree, H.24, flowers in December

MORACEAE

115. *Broussonetia greveana* (Baill.) C.C. Berg, Tree, VORY, A.57, sterile in May

OCHNACEAE

116. *Diporidium greveanum* v. Tiegh., TSIVANGO, Tree, CD.2, and H.36, flowers in November and December; H.45, fruits in December
117. *Diporidium pervilleanum* v. Tiegh, BORAMENA, TSIVANGO, Tree, JJP.8 and JJP.11 and H.35, flowers in November and December; H.12 and H.46, fruits in December; A.10, sterile in March
118. SAVONERA, A.43, sterile in April

OLEACEAE

119. *Noronhia boinensis* Perr., Tree, H.50, flowers in January; *Noronhia cf. boinensis,* TSILAITSY, A.79, fruits in June

ORCHIDACEAE

120. *Microcoelia perrieri* (Schltr.) Summerh., TINORINKIBO, A.134, sterile in June
121. *Oeceoclades decaryanus* (Perr.) Garay, H.51, flowers in January
122. *Sobennikoffia robusta* Schltr., likely to be present
123. *Vanilla cf. montagnacii* Portères, VAHIMALO, Liana, A.78, sterile in June

PAPILIONACEAE

124. *Chadsia flammea* Boj., Shrub, H.2, H.31 and JJP.4, flowers in November and December; *C. aff. flammea,* REMOTY, A.94, fruits in May
125. *Chadsia versicolor* Boj., Shrub, H.32, flowers in December; *C. aff. versicolor,* A.96, fruits in June
126. *cf. Dalbergia bracteolata* Bak., VAHINTA, A.22, sterile in April; H.56, sterile in January
127. *Dalbergia greveana* Baill., MANARIBAOMBY, Tree, A.41, sterile in April
128. *Dalbergia aff. purpurescens* Baill., MANARIBERAVY, Tree, A.56, fruits in May; CD.3, sterile in November
129. *Dalbergia chlorocarpa* Vig., MANARIVOLOE, Tree, A.90, sterile in May
130. *Dalbergia* sp., MANARY, Tree, A.6, sterile in April
131. *Pongamiopsis pervilleana* (Baill.) Vig. ex Peltier, HAZOMBOATANGO, Tree, A.47, sterile in May

PEDALIACEAE

132. *Uncarina* sp., FARAHETSY, Tree, A.155, sterile in May

RHAMNACEAE

133. *Berchemia discolor* (Klotzsch.) Hemsley, TSIANDALA, Tree, A.7, sterile in March

134. *Colubrina alluaudii* (Perr.) Cap., Tree, A.58, fruits in May
135. *Colubrina decipiens* (Baill.) Cap., TRATRAMBORONDREO, Tree, A.144, fruits in May
136. *Gouania lineata* Tul., VAHINDRANOMBAVANAOMBY, Liana, JJP.32, flowers in March; A.115, fruits in June

RUBIACEAE

137. *Breonia* sp., MOLOMPANGADIBERAVINA, Tree, A.53, fruits in April
138. *Carphalea* sp., MENAVAONY, Tree, A.158 and JJP.31, fruits in March; flowers in December
139. *Enterospermum* sp., TAOLANKENA, Shrub, A.65, fruits in June
140. *Hymenodictyon cf. louhavate* Hom., BEHOLITSE, Tree, A.19, sterile in April
141. *Poederia* sp., VAHIMANTSINA, Liana, A.91, fruits in June
142. *Pyrostria* sp., FATIKAHITRA, Tree, JJP.6, flowers in November; A.34, fruits in April
143. FATIKAHITRIBERAVY, A.81, sterile in June
144. FATIKAHITRIMADINIZAVY, Shrub, A.114, sterile in June
145. KOSOMPA, Shrub, H.49, flowers in January; JJP.23, fruits in March
146. LAMBO, Shrub, A.111, fruits in June
147. PAPOLAHY, Tree, A.88, fruits in May
148. TSIHATAIMBALY, Tree, A.120, sterile in June
149. TSIVOANINO, JJP.24, flowers and fruits in March
150. VOAFOTAKA, Tree, A.106, fruits in March
151. H.34, flowers in December
152. A.50, sterile in April
153. A.93, sterile in June
154. *Zanthoxylum tsihanimposa* Perr., MONONGO, Tree, H.60, sterile in January; HELABALALA, A.67, sterile in June

SAPINDACEAE

155. *Allophyllus cobbe* (L.) Raeusch., HAZOMPOSY, Tree, JJP.27, fruits in March
156. *Doratoxylon* sp., BANJONY, A.121, sterile in June
157. *Majidea zangabarica* Kirk. var. *madagascariensis* (Baill.) Cap., Tree, A.153, flowers in May

SAPOTACEAE

158. *Capurodendron rubrocostatum* (Jum. and Perr.) Aubr., NATOKORONONO, Tree, A.139, sterile in June

SOLANACEAE

159. *Solanum* sp., VOANGIVY, Shrub, H.40, flowers in December; A.84, flowers and fruits in June

STERCULIACEAE

160. *Byttneria oligantha* J. Ar., VAHINAMALONA, Liana, JJP.9bis and H.22, flowers in November and December; A.76, fruits in May
161. *Dombeya* sp., Shrub, H.38, flowers in December

(*Continued*)

162. *Hildegardia erythrosyphon* (Baill.) Kost., Vonoa, Tree, A.51, sterile in April
163. *Nesogordonia cf. ambalalensis* J. Ar., REFEKO, Tree, A.71, sterile in March
164. *Nesogordonia cf. micrantha* J. Ar., HAZOMENARANO, Tree, A.151, sterile in June

TACCACEAE

165. *Tacca leontopetaloides* (L.) Kuntze, TAVOLY, H.57, sterile in January

TILIACEAE

166. *Grewia cyclea* Baill., SELIBE, Tree, H.19, flower buds in December; H.10 and JJP.22, young fruits in December; A.32, fruits in April
167. *Grewia dombeyopsis* Cap., SELIMBIVY, H.43, flowers in December; A.15, sterile in April
168. *Grewia glandulosa* Vahl, Tree, H.52, fruits in January
169. *Grewia triflora* (Boj.) Walp., SELY, Tree, A.68, fruits in May
170. *Grewia voloina* Cap., SELIMALIMATSY, Tree, A.12, sterile in April
171. *Grewia* sp., SELIFOTIVORY, A.27, sterile in April
172. *Grewia* sp., SELIMBITIKA, JJP.25, fruits in March
173. *Grewia* sp., VALO, Tree, H.9, flowers in December; A.11, sterile in April
174. *Grewia* sp., JJP. 7, flowers in December
175. SELIMAVO, Tree, A.70, fruits in May
176. A.140, Tree, sterile in May

TURNERACEAE

177. *Piriqueta integrifolia* (Clav.) Cap., JJP.10 and H.30, Tree, flowers in November and December

VERBENACEAE

178. *Clerodendrum involucratum* Vatke, MAVOALA, Shrub, A.105, flowers in June
179. *Vitex menabeensis* Cap., Shrub, JJP.19, flowers in November; H.21, flowers and fruits in December

VITACEAE

180. *Cyphostemma pachypus* Desc., VOALOMBOKALA, Liana, A.49, sterile in May
181. A.77, Liana, sterile in June

SPECIES UNKNOWN

182. BETONDRO, A.89
183. HAZOFIA, A.63
184. HAZOMENALALA, A.3
185. HAZONDANDIALA, A.95
186. LALIANGIALA, A.37
187. NATO, A.103
188. PITIKALA, A.38
189. SOMPATSALA, A.29
190. TAINAKANGA, A.126
191. TSIENGENA, A.59
192. TSILOMAY, A.92
193. TSIVANINO, A.69

and numbers: H.6, A.62, A.74, A.101, A.123, A.124, A.125, A.131, A.137, A.148, A.154

APPENDIX I (*Continued*)

SPECIES SUPPOSED TO BE PRESENT (no herbarium specimens)

Secamopsis madagascariensis Boj., ASCLEPIADACEAE
Crateva greveana Baill., CAPPARIDACEAE
Rhopalocarpus lucidus Boj., RHOPALOCARPACEAE

[a] A. = J.-P. Abraham; CD. = P. Charles-Dominique; H. = C.M. Hladik; EP. = E. Pages; GP. = G. Pariente; JJP. = J.-J. Petter.

APPENDIX II

Checklist of the Vernacular Names of the Plant Species of Marosalaza Forest [a]

Anakaraky (36)	Hazomenalala (184)	Mantaora (101)
Anatsiko (79)	Hazomenarano (164)	Maronono (43)
Andriambolafoty (67)	Hazomipika (9)	Matambelo (39)
Antso (73)	Hazompody (106)	Mavoala (178)
Amaniomby (50)	Hazomposy (155)	Menahihy (61)
Babo (54)	Hazondandiala (185)	Menambahy (90)
Banaky (33)	Hazondolo (1)	Menavahony (138)
Banjony (156)	Hazonta (88)	Miateloravy (71)
Beholitse (140)	Helabalala = monongo (154)	Mita (53)
Betondro (182)	Kaboka (8)	Molompangadibera-vinamadinidravina (137)
Bokatsifaka (108)	Karimbelalahy (57)	Monongo (154)
Boramena (117)	Kililo (52)	Nato (187)
Boy = malamasefoy (38)	Kily (40)	Natokoronono (158)
Fangoky (37)	Kitata (64)	Papolahy (147)
Farafatsy (75)	Kitatamalandy (94)	Piripitsokala (60)
Farahetsy (132)	Kofafa (47)	Pitikala (188)
Fatiky (109)	Kompotsoy (13)	Reampy (45)
Fatikahitra (142)	Kosompa (145)	Refeko (163)
Fatikahitriberavy (143)	Lalangiala (186)	Relefo (55)
Fatikahitrimadi-nizavy (144)	Lambo (146)	Remoty (124)
Halampo (97)	Lamoty (83)	Reniala (20 and 21)
Halomboro (112)	Lombiry (11)	Roy (111)
Handy (104)	Mainaty (96)	Sakoambanditsy (7)
Harofiboy (26)	Malamasefoy (38)	Sanira (77)
Harofifotsy (27)	Mamiky (72)	Sarongaza (35)
Harofimena (25)	Mampandra (102)	Sarongazala (34)
Harofy (24)	Manandry (78)	Savonera (118)
Harongampanihy (89)	Manaribaomby (127)	Sekatsy (28)
Hasikitoky (95)	Manariberavy (128)	Selibe (166)
Hazoboenga (59)	Manarivoloe (129)	Selifotivory (171)
Hazofia (183)	Manary (130)	Selimalitsy (170)
Hazomalany (87)	Mandakolahy (103)	Selimavo (175)
Hazombo (84)	Mandifika (76)	Selimbitika (172)
Hazomboatango (131)	Mandravasaretra (58)	Selimbivy (167)
Hazomby (93)	Mangarahara (19)	Sely (169)
Hazomena (80)	Manjakabetany (31)	

(*Continued*)

APPENDIX II *(Continued)*

Soahy (65)	Tsidikandambolahy (107)	Vahindranambava-nahomby (136)
Sofikomba (23)	Tsiengena (191)	Vahinta (126)
Sofitamba (92)	Tsihataimbaly (148)	Vahironto (14)
Somangibe (42)	Tsilaitsy (119)	Valo (173)
Sompatsala (189)	Tsilavondria (81)	Vaotompaosa (6)
Tahila (100)	Tsilomay (192)	Vaovy (41)
Tainakanga (190)	Tsimaty (91)	Velatra (2)
Talamena (30)	Tsivango (116)	Velatrataka (3)
Talinala (48)	Tsivanino (193)	Voafotaka (150)
Taly (49)	Tsivoanino (149)	Voahavory (12)
Tambakonjiriky (51)	Tsivoaninombezo (29)	Voalombokala (180)
Taolankena (139)	Tsivoanizao (44)	Voangivy (159)
Tavoly (165)	Tsomontsoy (17)	Voantsilana (10)
Tinorinkibo (120)	Vahimantsina (141)	Volafoty (66)
Tohiravy (18)	Vahimalo (123)	Volily (5)
Tratramboron-dreo (135)	Vahimbaraongo (70)	Vonoa (162)
Tsiandala (133)	Vahinamalona (169)	Vory (115)
Tsibosibosy (86)		

[a] The numbers refer to species listed in Appendix I. Most of the names in Sakalava language are from J.-P. Abraham.

REFERENCES

Albignac, R. (1976). L'écologie de *Mungotictis decemlineatas* dans les forêts décidues de l'Ouest de Madagascar. *Terre Vie* **30**, 347–376.

Anonymous (1972). "Inventaire botanique et biologique de certains peuplements de la forêt naturelle des Côtes Orientales et Occidentales à Madagascar," Doc. de travail F.A.O., FO: SF/MAG. Tananarive.

Aubreville, A. (1957). Accord à Yangambi sur la nomenclature des types africains de végétation. *Bois For. Trop.* **51**, 23–27.

Baker, H.G. (1978). Chemical aspects of the pollination biology of woody plants in the tropics. *In* "Tropical Trees as Living Systems" (P.B. Tomlinson and M.H. Zimmermann, eds.), pp. 57–82. Cambridge Univ. Press, London and New York.

Baker, H.G., and Baker, I. (1975). Studies of Nectar-constitution and pollinator -plant coevolution. *In* "Coevolution of Animals and Plants" (L.E. Gilbert and P.H. Raven, eds.), pp. 100–140. Texas Univ. Press, Austin.

Bearder, S.K., and Doyle, G.A. (1974). Ecology of bushbabies, *Galago senegalensis* and *Galago crassicaudatus,* with some notes on their behavior in the fields. *In* "Prosimian Biology" (R.D. Martin, G.A. Doyle, and A.C. Walker, eds.), pp. 109–130. Duckworth, London.

Bourlière, F., and Équipe PBI/Sahel (1978). La savane sahélienne de Fété Olé, Sénégal. *In* "Ecosystèmes terrestres" (M. Lamotte and F. Bourlière, eds.), pp. 187–229. Masson, Paris.

Charles-Dominique, P., and Hladik, C.M. (1971). Le Lépilemur du sud de Madagascar: Ecologie, alimentation et vie sociale. *Terre Vie* **25**, 3–66.

Debray, M., Jacquemin, A., and Razafindrambao, R. (1971). Contribution à l'inventaire des plantes médicinales de Madagascar. *Trav. Doc. ORSTOM,* No. 8.

Frankie, G.W., Baker, H.G., and Opler, P.A. (1974). Comparative phenological studies of trees in tropical wet and dry forests in the lowlands of Costa Rica. *J. Ecol.* **62,** 881–919.

Gachet, C. (1971). Premiers résultats d'une période d'observations phénologiques à Marofandilla-Morondava. *C.T.F.T. Madagascar, Sol Foret* **51.**

Galat, G., and Galat-Luong, A. (1977). Démographie et régime alimentaire d'une troupe de *Cercopithecus aethiops sabaeus* en habitat marginal au Nord Senegal. *Terre Vie* **31,** 557–577.

Golley, F.B., and Medina, E. (1975). "Tropical Ecological Systems. Trends in Terrestrial and Aquatic Research." Springer-Verlag, Berlin and New York.

Guillaumet, J.L., and Koechlin, J. (1971). Contribution à la définition des types de végétation dans les régions tropicales (exemple de Madagascar). *Candollea* **26,** 263–277.

Hladik, A. (1974). Importance des lianes dans la production foliaire de la forêt équatoriale du Nord-Est du Gabon. *C.R. Hebd. Seances Acad. Sci. Ser. D* **278,** 2527–2530.

Hladik, A. (1978). Phenology of leaf production in a rain forest of Gabon: Distribution and composition of food for folivores. *In* "The Ecology of Arboreal Folivores" (G.G. Montgomery, ed.), pp. 51–71. Smithson. Inst. Press, Washington, D.C.

Hladik, A., and Hallé, N. (1979). Note sur les endocarpes de quatre espèces de *Spondias* d'Amérique (Anacardiacées). *Adansonia,* sér. 2, **18,** 487–492.

Hladik, A., and Hladik, C.M. (1969). Rapports trophiques entre végétation et Primates dans la forêt de Barro Colorado, Panama. *Terre Vie* **1,** 25–117.

Hladik, A., and Hladik, C.M. (1977). Signification écologique des teneurs en alcaloïdes des végétaux de la forêt dense: Résultats des tests préliminaires effectués au Gabon. *Terre Vie* **31,** 515–555.

Hladik, C.M. (1977). A comparative study of the feeding strategies of two sympatric species of leaf monkeys. *In* "Primate Ecology" (T.H. Clutton-Brock, ed.), pp. 323–353. Academic Press, New York.

Hladik, C.M. (1978). Adaptive strategies of primates in relation to leaf-eating. *In* "The Ecology of Arboreal Folivores" (G.G. Montgomery, ed.), pp. 373–395. Smithson. Inst. Press, Washington, D.C.

Hladik, C.M., and Hladik, A. (1967). Observations sur le rôle des primates dans la dissémination des végétaux de la forêt gabonaise. *Biol. Gabonica* **3,** 43–58.

Hladik, C.M., and Hladik, A. (1972). Disponibilités alimentaires et domaines vitaux des primates à Ceylan. *Terre Vie* **26,** 149–215.

Humbert, H. (1965). Description des types de végétation. *In* Notice de la carte internationale du tapis végétal et des conditions écologiques au 1/1.000.000—Madagascar. *Trav. Sect. Sci. Tech., Inst. Fr. Pondichéry* **6,** 46–78.

Humbert, H., and Cours Darne, G. (1965). Notice de la carte international du tapis végétal et des conditions écologiques au 1/1.000.000—Madagascar. *Trav. Sect. Sci. Tech., Inst. Fr. Pondichéry* **6,** 1–162.

Janzen, D.H. (1967). Synchronization of sexual reproduction of trees within the dry season in Central America. *Evolution* **21,** 620–637.

Janzen, D.H. (1969). Seed eaters versus seed size, number, toxicity and dispersal. *Evolution* **23,** 1–27.

Jolly, A. (1966). "Lemur Behavior. A Madagascar Field Study." Univ. of Chicago Press.

Koechlin, J., Guillaumet, J.L., and Morat, P. (1974). Flore et végétation de Madagascar. *Flora Veg. Mundi* **5,** 1–687.

Lamotte, M., and Bourlière, F. (1978). "Problèmes d'écologie: Structure et fonctionnement des écosystèmes terrestres." Masson, Paris.

Legris, P., and Blasco, F. (1965). Carton des bioclimats. *In* Notice de la carte internationale du tapis végétal et des conditions écologiques au 1/1.000.000—Madagascar. *Trav. Sect. Sci. Tech. Inst. Fr. Pondichéry* **6,** 31-45.

Leigh, E.G., and Smythe, N. (1978). Leaf production, leaf consumption, and the regulation of folivory on Barro Colorado Island. *In* "The Ecology of Arboreal Folivores" (G.G. Montgomery, ed.), pp. 33-50. Smithson. Inst. Press, Washington, D.C.

Mueller-Dombois, D. (1968). Ecogeographic analysis of a climate map of Ceylon with particular reference to vegetation. *Ceylon For.* **8,** 39-58.

Perrier de la Bathie, H. (1921). La végétation malgache. *Ann. Musée Col. Marseille* **9,** 1-268.

Petter, J.J., Schilling, A., and Pariente, G. (1971). Observations éco-éthologiques sur deux lémuriens malgaches nocturnes: *Phaner furcifer* et *Microcebus coquereli. Terre Vie* **13,** 287-327.

Ridley, H.N. (1930). "The Dispersal of Plants throughout the World." Reeve & Co., Ashford, Kent.

Rudran, R. (1978). Intergroup dietary differences and folivorous tendencies of two groups of blue monkeys, *Cercopithecus mitis. In* "Ecology of Arboreal Folivores" (G.G. Montgomery, ed.), pp. 483-503. Smithson. Inst. Press, Washington, D.C.

Sussman, R. W. (1978). Nectar-feeding by Prosimians and its evolutionary and ecological implications. *In* "Recent Advances in Primatology" (D.J. Chivers and K.A. Joysey, eds.), Vol. 3, pp. 119-125. Academic Press, New York.

Wiens, D., and Rourke, J.P. (1978). Rodent pollination in Southern Africa *Protea spp. Nature (London)* **276,** 71-73.

2 Feeding Strategies of Five Nocturnal Prosimians in the Dry Forest of the West Coast of Madagascar

C. M. HLADIK, P. CHARLES-DOMINIQUE, and J. J. PETTER

INTRODUCTION

The adaptive strategy of animal populations allows a species increased fitness in response to important variations of environmental conditions mainly affecting food availability. In evolutionary terms, some species invest in the efficiency of reproductive functions for maximum population growth when conditions permit (r strategy), while other species maintain a fairly constant population (K strategy) through maximum investment in individual fitness (MacArthur and Wilson, 1967). The predictability of food production is a determinant parameter of such mechanisms, the K strategy existing only in the presence of regular cycles of overall food production. By contrast, species adaptation to food resources described as "generalization" and "specialization" (MacArthur, 1965) may or may not correspond to the temporal patterning of food availability. For instance, specialized species such as most folivorous mammals are adapted to a set of predictable food resources (A. Hladik, 1978), while other specialists, mainly among monophagous invertebrates, evolved towards r strategies to maximize the output from locally abundant resources unevenly distributed in space and time (Wilson, 1975).

Among primates, feeding strategies which have been observed in different species are intermediate cases of specialization concerning differential utilization of food resources. The most specialized is a "low-cost and low-energy return" strategy of some folivores (Hladik,1977a) which allows large populations to subsist on the most abundant but least nutritious plants. The most generalized is the "high-cost and high-energy return" strategy of many Cercopithecinae and Chimpanzees (Hladik, 1977b). Both tendencies are obviously the expression of the K strategy, and tend to maintain constant population densities (Charles-Dominique and Hladik, 1971). Although primate population ecology suffers from the small size of most samples observed, the accuracy of a multitude of field studies has allowed detailed analysis of feeding strategies in relation to physiology, behavior, and sociology, which support these views (C. M. Hladik and Chivers, 1978).

A comparison of the different adaptations of the nocturnal prosimian species of the Marosalaza forest (West Coast of Madagascar) provides additional information of this kind. In this region, the environment presents a highly variable cycle (Chapter 1), with predictability of food supply only possible on a long-term basis. The prosimian species have adopted several ecophysiological or behavioral solutions, a brief account of which has already been presented by Petter (1978). Two of these species, *Phaner furcifer* and *Microcebus coquereli,* are the main subject of other chapters included in this volume, by Charles-Dominique and Petter, and by Pages, respectively. A third species, *Lepilemur ruficaudatus,* can be compared to other Sportive Lemurs studied in detail elsewhere (C. M. Hladik and Charles-Dominique, 1974). Accordingly, in the present comparative study, we mainly develop detailed information concerning the two Cheirogaleinae, *Cheirogaleus medius* and *Microcebus murinus,* which live in sympatry with the three other nocturnal species in the Marosalaza forest.

This chapter is based on data obtained during successive and partly overlapping field studies of the authors (J. J. Petter, October/November, 1973 and January/February, 1974; C. M. Hladik, December, 1973/January, 1974; P. Charles-Dominique, November/December, 1973 and May/June, 1974 and is completed by field observations of other participants of the Morondava Programme (G. F. Pariente and E. Pages). The interpretation of most results was permitted by careful analysis of physiological cycles of the different prosimians, conducted simultaneously and following completion of the field work, in the animal house of the Brunoy Laboratory (see Chapters 2, 6, and 7).

The different techniques for recording data on plant production and for animal trapping and marking, which are also common to other papers in this volume, will be briefly described in the relevant sections.

SEASONAL VARIATIONS IN FOOD PRODUCTION AND FOOD COMPOSITION IN THE MAROSALAZA FOREST

The environmental conditions in the Marosalaza forest are described in Chapter 1. Plant diversity is not as important as in most tropical forests inhabited by primates. Nevertheless, different life cycles of tree and liana species provide a successional emergence of various food resources.

Yearly variations, in terms of the number of plant species bearing flowers, fruit, and new leaves (Fig. 1), provide only an approximate picture of the actual changes of food availability in the Marosalaza forest, since the biomass of the different species varies considerably. Accordingly, an index was calculated which takes into account the total basal area of the most common species present in the sample and reflects the total biomass. The values obtained by adding these "biomass indices" yielded overall variations roughly correlated with species abundance: a peak for flowers in November, but a slightly more even distribution of fruit production between December and June. Since these different values are not significant for plant species present in small quantity, potential food availability could not be recorded in terms of absolute production, except during a limited period (see Table V, Chapter 1). Nevertheless, the different methods of approach yielded sufficiently similar results in terms of relative variation of potential food production to consider Fig 1. as a valid base for comparing the adaptations of different prosimian species.

Seasonal Variations in Flowers

There is a very marked period of flowering in October/November preceding the rainy season, which lasts for only a short time after the rain

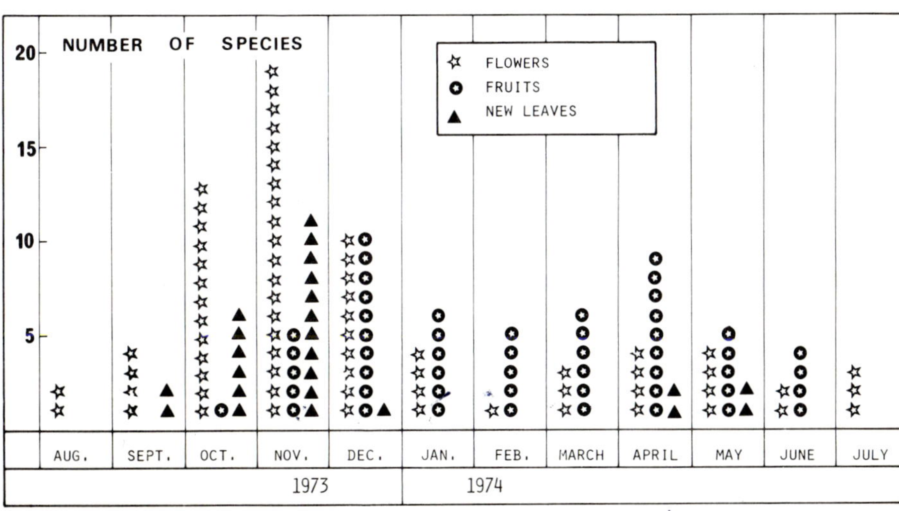

Fig. 1. Variation of food availability throughout the year, in the Marosalaza forest, according to the number of species producing potential foods in a sample of 2200 m² (J.- P. Abraham's transect analyzed by A. Hladik, Chapter 1, Table I).

Fig. 2. The Boy flower, *Delonix floribunda* (A), in which was found the most abundant production of nectar. The detail (B) shows the unique central petal totally filled with 0.5 gm of nectar (species 38 in Appendix I of Chapter 1; photos by C. M. Hladik).

starts. Many species of Marosalaza forest plants bear flowers with a very large quantity of nectar, such as the Boy, *Delonix floribunda* (28) * (Fig.2). Sussman (1978) already noted the importance of some large flowers in the diet of diurnal and crepuscular prosimians, which are major pollinators of these plants. In Marosalaza, the sudden abundance of such potential foods during the austral spring, corresponds to changes in physiological cycles of activity, especially in *Microcebus murinus* and *Cheirogaleus medius* (Chapter 6). During the dry season, only a limited number of plants produce flowers, but the large Baobabs, *Adansonia* spp. (20/21) might provide an important supply for still active species such as *Microcebus coquereli*.

Seasonal Variations in Fruits and Young Leaves

Fruits are available throughout the rainy season and until June. In December, the maximum production of fruits such as the Malagasy plum, *Flacourtia indica* (83), the fruit of Iron wood, *Strychnos decussata* (93), the Forest grape, *Cyphostemma pachypus* (180), a bottle-like vine, (see Tables

* The numbers in parentheses refer to Appendix I of Chapter 1.

I and II, Chapter1), occurs simultaneously with the decreased availability of nectar and other edible parts of flowers produced by different species. Fruits are also abundant in February/March, when the young of different prosimian species start feeding independently. Fruit production which is maintained until the beginning of the dry season, in April/May, permits species such as *Cheirogaleus medius* and *Microcebus murinus* to accumulate fat reserves preceding the period of lethargy (Chapter 6).

Young leaves are also an important potential food resource (A. Hladik, 1978) and are partly used by Cheirogaleinae before fruits become abundant. By contrast, tough mature leaves are potential food for *Lepilemur ruficaudatus* only. The major importance of leaf production is the direct impact on invertebrate production (see below) which is of paramount importance for most species of nocturnal prosimians.

Composition of Plant Parts

Samples of prosimian foods were collected in the Marosalaza forest, and preserved for analysis according to standard techniques (Hladik, 1977c). Results of the analysis of the major types of food are presented in Table I.

Since flowers and fruits are likely to provide the staple food of various prosimian species during long periods, their relative composition of carbohydrates and protein might be determinant of the dietary balance. In this respect, it is remarkable that the mean protein content of the different fruit samples (7.3%) is high compared to that of the fleshy parts of most fruits of the African rain forest and the Asian semi-deciduous forest (rarely in excess of 5%; Hladik, 1977b; 1978a). For animals feeding on insects and fruits, the risk of facing protein deficiency during periods of insect scarcity is thus lower than in other environments. For example, the protein-rich fruit of *Operculicarya gummifera* (7) which is also rich in fat, is a highly valuable food for *Cheirogaleus medius* at the beginning of its period of activity. This animal feeds on the pulp of such fruits and disperses the seeds in feces, and consequently exerts a selective pressure in favor of the most attractive fruits in terms of nutrient. Although this is a general phenomenon (see discussion in Chapter 1), selection pressures are influenced by the long dry season in the Marosalaza forest, which results in a decrease of most food resources (especially protein, i.e., insects; see below), and explains the unusually high protein content of fruits (Table I).

Gums and tree exudates have a very low protein content which implies the necessity of a protein supplement (insect) for species feeding on these resources. One sample of gum of *Terminalia tricristata* (49) presents a high content of soluble sugar extracted after weak hydrolysis. This appears surprising, because in other samples of gum (mainly those of lianas and trees of Gabon), the same type of hydrolysis does not break the apparently

TABLE I

Composition of Samples of Foods Eaten by Prosimians[a]

Type of food	Percentage dry weight: Protein ($N \times 6.25$)	Percentage dry weight: Reducing sugars (after weak hydrolysis)
Flower extract		
Delonyx floribunda H. 14 A (38)	6.4	68.2
Fruits (fleshy parts only)		
Operculicarya gummifera H. 3 A (7)	9.5	14.2
Physena sessiliflora H. 53 A (85)	8.3	32.5
Strychnos decussata H. 25 A (93)	7.1	55.9
Grewia glandulosa H. 52 1 (168)	6.0	15.8
Evonymus plurostyloïdes GP. A (43)	5.8	17.1
Gums		
Terminalia mantaliopsis PCD. 3 F (48)	5.2	
Terminalia tricristata H. 58 F (49)	3.3	
Terminalia tricristata EP. 3 S (49)	2.6	87.8
Other tree exudate		
Sap of *Adansonia fony* PCD. 2 F (20)	5.1	
Insect secretion		
Dried secretion from larval colony of *Flatida coccinea* EP. 1 S	1.4	5.0

[a] Collected in the Marosalaza forest by P. Charles-Dominique (PCD.), C. M. Hladik (H.), E. Pages (EP.), and G. F. Pariente (GP.). The reference numbers in parentheses are those in Appendix I in Chapter 1.

strong chains of pentoses (C. Mercier, personal communication, and unpublished data of C. M. Hladik and Charles-Dominique), and in these cases, only a small amount of soluble sugars is obtained. This type of analysis of soluble sugars has been chosen (Hladik *et al.*, 1971) to account for what is likely to be assimilable by most mammals without a particular digestive adaptation. Accordingly, some of the gums available in the Marosalaza forest appear to be readily digestible by nonspecialized prosimians. The specialized gum eaters might thus have originated from this type of forest; but further investigation should be necessary to confirm this hypothesis.

The investigations concerning the possible toxicity of plants in the Marosalaza forest have been limited to the alkaloid screening of Hladik (Chapter 1). Other secondary compounds, such as tannins, glycosides, and saponines, may also be toxic (Whittaker and Feeny, 1971) and were actually found in some of the samples from the West Coast of Madagascar analyzed by Debray *et al.* (1971). The overall results of the alkaloid survey indicate concentration of alkaloids mainly in evergreen species. This may reduce the possible food choices for a folivore such as *Lepilemur ruficaudatus* during the dry season. In contrast, the other prosimian species feeding on fruits and nectars are not expected to face plant toxicity.

Invertebrate Production

Sampling invertebrate production in order to determine what is available as food to prosimians, necessarily implies a bias due to the method of collection. An ultra violet lamp located near the forest border was used to attract flying insects in a trap, during 2 hours at the beginning of each night. After the initial period of intense capture (> 25 gm dry weight, per night), the "production," including mainly Coleoptera, was more stable and probably reflected variations of a part of the invertebrate food available to nocturnal prosimians. The amount of insects progressively increased from January to February (from 6 to 15 gm), and subsequently tended to decrease (about 8 gm in March). This insect population thus followed the pattern of leafing of most trees illustrated by Fig. 1, the insects being available about 2 months after the leaves have developed.

A more reliable estimate is provided by caterpillars' feces collected and weighed in the litter (Table V, Chapter 1). The increase of caterpillar feeding activity is also subsequent to leaf growth with a maximum reached in May. It thus can be assumed that insect populations attain maximum density at the end of the rainy season.

Accordingly, the food available to most nocturnal prosimians, which is mainly composed of carbohydrates (nectars) in November, and includes fruits with a fairly high protein content in December/January, would progressively become richer in insect protein in February/March (insect composition includes approximately 70% protein, dry weight; Hladik, 1977b). The continuation of fruit production in March/April (Fig. 1) would subsequently result in a diet including more carbohydrates for the species which increase food intake before the dry season.

Invertebrate availability was not homogeneous in the study area, especially during the dry season. Different samples collected by E. Pages showed that the amount of arthropods was twice as large in the dense vegetation around the temporary pond as in the dryer parts of the forest (Table II). The distribution of invertebrate populations is also influenced by forest structure, for example, Orthoptera are present in larger quantity in Area 3. An extrapolation from these results and leaf production indicates an order of magnitude of 0.5 to 1.0 kg per hectare of invertebrate biomass.

TABLE II

Comparison of Arthropod Samples Classified According to Possible Level of Identification, in Different Areas of the Marosalaza Forest[a]

	Number of individuals (percentage occurrence)	Dry weight (gm)	Percentage of dry weight
Area 1			
Arachnids	55(53)	0.16	30
Blattaria (cockroaches)	24(23)	0.12	23
Mantiidae	7(6)	0.08	15
Coleoptera	5(5)	0.04	8
Hemiptera	4(4)	0.04	8
Homoptera	4(4)	0.03	6
Lepidoptera	2(2)	0.03	6
Orthoptera (crickets)	1(1)	0.01	2
Hymenoptera	1(1)	0.01	2
Area 2			
Arachnids	56(55)	0.21	41
Blattaria	22(21)	0.08	15
Orthopthera	5(5)	0.05	10
Lepidoptera	3(3)	0.05	10
Homoptera	7(7)	0.04	8
Mantiidae	4(4)	0.04	8
Misc. including ants	5(5)	0.04	8
Area 3			
Blattaria	38(26)	0.28	27
Arachnids	74(50)	0.26	25
Orthopthera	8(5)	0.22	21
Hemiptera	13(9)	0.12	12
Lepidoptera	8(5)	0.08	8
Coleoptera	4(3)	0.04	4
Homoptera	5(3)	0.02	2
Hymenoptera	4(3)	0.01	1

[a] The samples were collected by E. Pages from the leaves of *Buxus madagascariensis* collected at random up to 5000 gm fresh weight. The three areas investigated (Area 1 along trails A to M; Area 2 along trails M to R; Area 3 along trails R to W) have also been compared for population density of *Microcebus coquereli* (see Chapter 4, Table I).

Other types of invertebrates are not directly preyed upon by prosimians but nevertheless play a role in producing secretions available as food. Different cochineals have been observed as well as homopterans. The most abundant species is *Flatida coccinea* (Homoptera, Fulgoridae; Fig. 3)* which forms large colonial aggregates on the vine *Elachyptera minimiflora*

* Identification was made by M. Boulard (Laboratoire d'Entomologie, Museum National d'Histoire Naturelle).

Fig. 3. The homopteran, *Flatida coccinea,* feeding on a stem of the vine *Elachyptera minimiflora* during austral summer. At the larval stage, during the dry austral winter, these insects secrete large quantities of honey dew available as food to *Microcebus coquereli* (photo by C. M. Hladik).

(88). During the dry season, the larval forms produce large droplets of honey dew which may be licked by *Microcebus coquereli* (see Chapter 4). This secretion falls on the leaves under the colony, and dries to form a white sweet concretion which remains edible. Results of analysis of this product are shown in Table I. The very low protein content is not surprising since the plant sap extracted by the insect is initially poor. In addition, the low value of reducing sugars is probably due to a nonreducing component which has been described as "mannite" (Decary, 1946) and was exploited by humans.

Vertebrates, mainly chameleons which appeared to be abundant in the Marosalaza forest, also form a part of the food available for prosimians.

POPULATIONS OF NOCTURNAL PROSIMIANS IN RELATION TO FOOD AVAILABILITY

Population censuses were conducted in 1973 and 1974, in the study area of the Marosalaza forest (see Fig. 5, Chapter 1). The animals were captured and consistently recaptured in spring-loaded traps, and marked with different combinations of ear cuts and complementary visible marks by depilation of some parts of the tail, according to the different methods of Charles-Dominique (1977). These data (Table III) are the most reliable, since the nocturnal observations conducted during several months showed that most of the animals present in the center of the study area consisted of marked individuals. In species for which home ranges have been measured (see Chapters 3 and 4), the densities indicated in Table III have been calculated according to these measures.

Other censuses have been made along transects, in order to obtain comparative data from different areas (Table IV). These were conducted repeatedly by two observers (C. M. Hladik and P. Charles-Dominique in November, 1973; C. M. Hladik and G. F. Pariente in January, 1974) walking slowly along trails with headlights, and using a measuring device ("Topofil") to locate accurately all observations. In a given area, successive surveys yielded very similar results.

Population densities of the different prosimian species have the same order of magnitude in different parts of the closed forest. These densities are extremely high, especially for *Cheirogaleus medius* and *Microcebus murinus* (Table III). One of the transects in the closed forest (Table IV) was located about 5 km south of the study area (a trail previously opened for oil prospection was reused). The lower densities calculated in this case for *C. medius* and *M. murinus* may well be due to the poor visibility in the undergrowth of the forest where these species generally forage. On the contrary, the population densities of *Phaner furcifer* and *Lepilemur*

TABLE III

Population Densities and Biomasses of Nocturnal Prosimians in the Study Area[a]

	Cheirogaleus medius	*Microcebus murinus*	*Microcebus coquereli*	*Phaner furcifer*	*Lepilemur ruficaudatus*
Number of animals:	31	72	9	14	4
Area controlled (ha)	(9)	(20)	(30)	(25)	(2)
Estimated population per km² (100 ha)	350	400	30	50–60	250
Biomass (gm/ha)	500–700	200	100	200	2000

[a] Marosalaza Forest, 50 km north of Morondava.

TABLE IV

Comparison of Population Densities in Two Different Areas of the Forest of the West Coast of Madagascar

	Cheirogaleus medius	*Microcebus murinus*	*Microcebus coquereli*	*Phaner furcifer*	*Lepilemur ruficaudatus*
Transect in closed forest (3 ha); Number of animals	6	2	1	3	10
Estimated population per km²	200	60	30	100	330
Trail from closed forest to Sira Sira (4.4 ha); Number of animals	1	6	0	2	2
Estimated population per km²	20	130	0	40	40

ruficaudatus are certainly exaggerated as a consequence of the excellent visibility in the tree tops where records were possible at distances exceeding 15 m from the trail (see methodological discussion about "visibility profiles" in Charles-Dominique and Hladik, 1971). Nevertheless, this method of census showed clear-cut differences between the closed forest and the transitional part on dry and/or salty soils towards the "Sira sira," where *Microcebus coquereli* is totally absent. The low population densities of the other species along this transect (Table IV) indicate a gradation in which most animals have a normal density near the tall forest, which diminishes fairly abruptly towards the Sira sira. The two exceptions are *M. murinus,* that we observed in low bushes up to the extreme limit of the Sira sira, and *P. furcifer* which probably can survive on a shrub producing gums, closely related to the "Talys." The local decrease in population densities can be related to the obvious scarcity of food resources in the lower less dense parts of the forest.

Biomasses and Food Availability

The biomasses of the different prosimian species, which have been calculated according to population densities (Table III), are the best "transducers" of the possible feeding strategies. The classical approach of any ecosystem through a pyramid of the biomasses (Odum and Odum, 1959) has been refined by a classification of primate species into "dietary grades" (Hladik, 1975; Hladik and Chivers, 1978) which are sorts of intermediate levels in this pyramid, and correspond to particular dietary types. These grades are: animalivore (grade 1) presenting a gradual frugivorous tendency towards an intermediate type (grade 2) in which protein balance is obtained from a mixture of animal prey and plant parts, and the folivore/frugivore type (grade 3), exclusively vegetarian. The different levels in these three grades can be expressed by the logarithm of the biomass (in grams per hectare), from 1.0 (grade 1 with biomass 10), up to 3.9 (grade 3 with biomass smaller than 10,000). The interdependence between biomasses and dietary types is due to a fairly homogeneous production of the different food types in the tropical forests throughout the world (see A. Hladik, 1978; C. M. Hladik, 1980).

According to Table III, most nocturnal prosimian species are in grade 2 *(Cheirogaleus medius:* 2.8; *Microcebus murinus:* 2.3; *Microcebus coquereli:* 2.0; *Phaner furcifer:* 2.3). Only *Lepilemur ruficaudatus* is in grade 3. Prosimian species must thus depend to a large extent on plant food resources (see below, and Fig. 10), but the system of sharing plant production might be delicate, since the average production (about 3 tons/ha/year in terms of dry litter fall; Chapter 1) is one of the smallest among tropical forests.

Furthermore, the Marosalaza forest not only supports the total biomass of 3100 gm per hectare of nocturnal prosimians, but an even larger biomass of diurnal lemurs as well. The two diurnal species, *Propithecus verreauxi* and *Lemur fulvus,* have been briefly observed. Probably due to recent hunting pressures these species seem less abundant as compared to the more accurate records of Jolly (1966), Richard (1974), and Sussman (1974), obtained in southern areas supporting types of deciduous forest comparable to Marosalaza. Nevertheless, *L. rufus* and *P. verreauxi* are sympatric with *Lemur catta* in these other areas, and we might expect at least a few changes in their feeding strategies, due to the absence of a potential competitor for food resources in Marosalaza. Tree and liana leaves and fruits were used to a large extent, for instance the genera *Landolphia* (8), *Cordyla* (36), *Grewia* (168–173) by *L. rufus,* and *Cynanchum* (12), *Diospyros* (55–60) by *P. verreauxi.*
Cynanchum (12), *Diospyros* (55–60) by *P. verreauxi.*

The total biomass of prosimians utilizing the plant production of the Marosalaza forest is thus very high (between 6,000 and 10,000 gm/ha). It has indeed the same order of magnitude as in other tropical forests affected by important seasonal changes, such as the gallery forest of the south of Madagascar (Jolly, 1966; Charles-Dominique and Hladik, 1971), or the Wilpattu forest in Sri Lanka (Eisenberg *et al.,* 1972; Hladik and Hladik, 1972). However, in Marosalaza, production is significantly smaller, and the seasonal variations in food production and composition appear more important than in these forests. The diversification of the feeding strategies of the different prosimian species seems to be a response to these environmental constraints.

ADAPTIVE STRATEGIES OF PROSIMIANS

Hibernation: The Fat-Tailed Dwarf Lemur, *Cheirogaleus medius*

The most characteristic feature of *Cheirogaleus medius* is the ability to hibernate for at least 6 consecutive months and, in some instances, up to 8 months of the dry season. During this period, the animals remain totally inactive. They retire inside deep holes of tree trunks where three to five individuals may be piled upon each other, separated by a kind of soft lodge made of wet decaying wood.

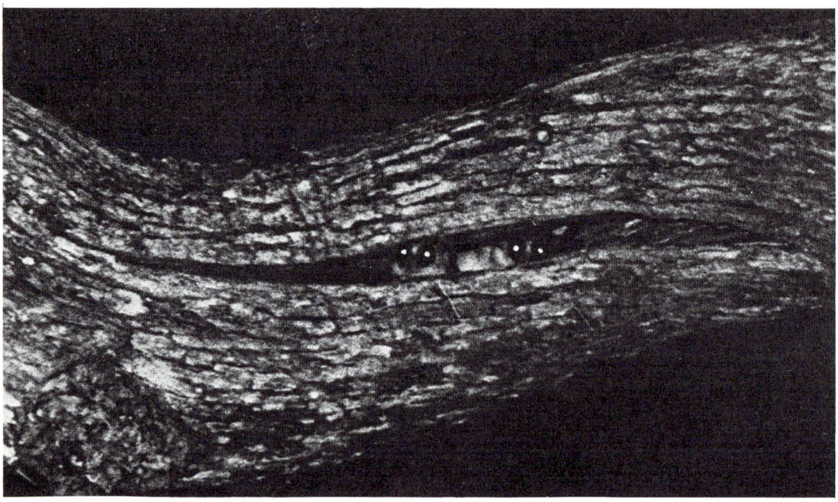

Fig. 4. A group of *Cheirogaleus medius* at the beginning of the rainy season, just emerging from the hollow trunk where they have been hibernating (photo by C. M. Hladik).

In the Marosalaza forest, the Fat-Tailed Dwarf Lemurs were first observed to emerge from their holes (Fig. 4) and forage at night at the end of November, just before the first rain started. Accordingly, as can be seen in Fig. 1, the animals are active exclusively during the period of intensive food production. Since populations of other nonhibernant prosimian species are limited by temporary reduction of food resources, *C. medius* can be considered as feeding mainly during the temporary period of *surplus* food.

Diet

The diet of *C. medius* was determined by direct observations at night, and from the analysis of fecal material collected when the animals were caught for marking. In this analysis, the baits used in the traps (bananas) were mixed with the other materials, but the remains of animal matter allowed a comparative study of the diets of different prosimian species captured in similar conditions.

Fruits formed the staple food (together with flowers in the beginning of the period of activity). Invertebrates were present in 50% of the feces, but in small amounts, the volume of chitin never exceding 10% of the other food remains. Coleoptera were the major type of prey (also observed: cockroaches, ants, cochineals, Cicada (?), and other undetermined invertebrates). The skin of chameleon was also found in one feces sample.

The small number of samples (31) did not permit following of the seasonal food cycle, but indicated an increase of invertebrate food in January, corresponding to the apparent increase of insect populations (see above). *C. medius* has been observed foraging for insects at medium height (4–8 m) in the Marosalaza forest.

The seeds of small fruits were intact in feces, mainly those of *Operculicarya gummifera* (7) and *Grewia glandulosa* (168), but the animals were also observed to eat pulps of larger fruits such as *Strychnos decussata* (93) and *Diospyros aculeata* (55), in December.

Flowers were licked and may form an important food supply. For instance, the nectar of the Boy flower, *Delonix floribunda* (28) (Fig. 2), seemed to be the unique food of one individual (♀8) followed by radio-tracking during 6 consecutive days in December. The animal moved from a nearby tree hole to the large Boy tree covered with flowers, and retired after a relatively short period of active feeding. Flowers of *Baudouina fluggeiformis* (31), a common small tree, were also frequently visited.

Some gums may be used (found in fecal samples) as well as many other plant or insect exudates (direct observations). When the animals licked leaves, it was difficult to identify the type of secretion (homopteran, cochineal, or nectar from flowers). Tender parts of plants may also be eaten in small amount.

In general, *C. medius* appears as an opportunistic frugivore complementing its diet with the most abundant and nutritious food types. The seasonal variation in diet follows food availability: nectars and fruits in November; fruits and an increasing proportion of insects in December/January/February. After this period, the proportion of fruits might be higher in the diet at the time of fattening preceding hibernation.

Variations in Body Weight

Capture, marking, and recapture of 31 *C. medius* permitted the recording of the overall variations in body weight in Marosalaza forest (Fig. 5). During the active period, the animals accumulate large fat reserves under the skin and inside the tail (Fig. 6). Adult mean body weight varied from 142 gm in November to 217 gm in March. The tail volume was measured by immersion (after wetting) in a graduated cylinder filled with water. Maximum adult tail volume (56 cm^3) was observed in May before hibernation, and minimum (9 cm^3) in November, after hibernation, the monthly means being, respectively, 42 cm^3 (SD = 9) and 15 cm^3 (SD = 4).

Similar body weight variations of *C. medius* have been subsequently recorded in artificial conditions (Chapter 6), but the period of "inactivity" was not as marked as in natural conditions (no true hibernation), and the

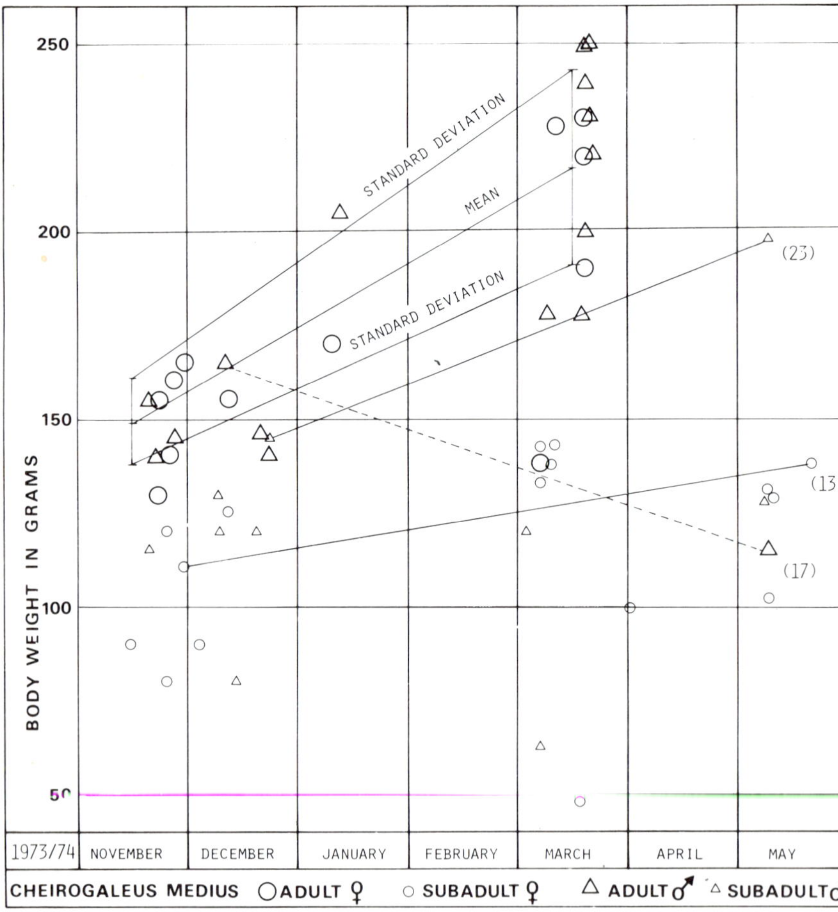

Fig. 5. Seasonal variation in body weight of *Cheirogaleus medius,* in the Marosalaza forest, according to data of captured and recaptured marked animals. The increase in adult body weight from November to March can be seen by the mean and standard deviation from 142 gm (SD = 12) to 217 gm (SD = 26). Note the important individual scatter in March, due to local conditions and/or individual status. The young animals (born in 1973/74) have the smallest weight. Individual data show growth of one subadult male (23), that of one subadult female (13), and that of an adult male (17) who lost weight (broken line) and was probably unable to survive the next dry season. Except for this last animal, only subadults were caught in May (adults began hibernating in March).

loss of weight less important. Nevertheless, these detailed records permitted the progressive changes in the patterns of food intake to be followed. There is a very marked increase of food intake concomitant with a decrease in activity preceding the period of lethargy. Furthermore the food choices were

Fig. 6. *Cheirogaleus medius* illustrating variations in tail volume. (A) The maximum size of tail, due to fat accumulation (photo by J. J. Petter). (B) The minimum tail volume, observed in November, just after hibernation in the Marosalaza forest (photo by A. Schilling).

oriented toward different food categories according to the period of the active cycle (Chapter 8).

These important variations in behavior and/or physiology can be interpreted according to the variations occurring in the wild, since they follow a similar overall pattern. At the beginning of the active period, when the food available was mainly carbohydrates (nectars and fruits), the animals tended to direct food choices toward this type of food (even when other types were available in artificial conditions). Dietary increase in protein which was due to insect availability in the Marosalaza forest also occurred in artificial standard conditions during the corresponding period of the annual cycle. Finally, the period preceding lethargy was marked by a diet rich in fruits for the animals in their natural environment, as well as for those studied in artificial standard conditions.

As a matter of fact, it appears that the feeding behavioral tendency of *C. medius* varies throughout the year and seems "preadapted" to the important changes of the natural environment (see discussion in Chapter 8).

Some other important changes in the apparent coefficient of food assimilation also precede the period of lethargy (Chapter 9).

The variations in body weight shown in Fig. 5 are thus related to successive adaptations in physiology and behavior, the most important being the sudden change in food choice, food intake, and overall activity, which precedes hibernation.

Reproduction

Mating was observed at the beginning of November, as soon as *C. medius* resumed activity after hibernation. Births occurred in January, at the time of increasing food availability.

Young animals have a relatively low body weight in March (Fig. 5) and continue to feed until May. At this time all the adults are already hibernating and thus do not compete for the remaining available food. The period of hibernation is consequently shorter for these young, and their loss in weight may be limited. Their lacteal last premolar is still present during this first hibernation, and is replaced only at the beginning of the next period of activity during which they attain the adult body weight.

Home Range

Most of the 31 marked *C. medius* did not move over ranges exceeding 200 m in diameter (about 4 ha). Larger displacement during the whole period of activity are generally by juveniles. Considering the very high population density (Table III), the ranges of the different adults must necessarily have important overlaps.

No detailed information concerning territoriality in this species could be obtained by direct observation during the wet austral summer. In captivity, individuals are generally intolerant to individuals of the same sex. The fecal marks which are very abundant, especially before the period of hibernation (Chapter 9), are likely to play a role when the animals resume activity.

Since no important movements were recorded at the beginning of the active period, except for two adults of small body weight (peripheral males?), it may be concluded that most adult *C. medius* stay in the same area during successive years.

The imperative necessity to find a shelter for hibernating seems as important as food resources for *C. medius*. It has been stressed (Chapter 1) that a remarkable aspect of the forest of the West Coast of Madagascar is the high frequency of hollow trunks in many common species such as the Anatsiko, *Securinega seyrigii* (79), the Harofy, *Commiphora* spp. (24/25), and the Iron wood, *Strychnos decussata* (93). This was a necessary condition permitting the emergence of the feeding strategy of *C. medius* which consists of harvesting "surplus" food when available, and withdrawing during the period of minimum production.

Reduction of Activity: The Lesser Mouse Lemur, *Microcebus murinus*

Microcebus murinus is not a true hibernant, but is comparable to *Cheirogaleus medius* by its period of lethargy (Chapter 6) which also occurs during the dry season. At this time, the animals may stay in hollow trunks for several consecutive days, and their overall activity is considerably reduced. Nevertheless, *M. murinus* are visible at night in the forest at any time of the year.

Feeding activity of the Lesser Mouse Lemur is thus maintained throughout the year, but decreases during the dry austral winter. At this time of reduced food availability, only small amounts are actually necessary, as was the case for the juvenile *C. medius,* the active period of which is longer than that of adults. The major proportion of annual food intake of *M. murinus* coincides with maximum fruit and insect production (austral spring and summer). Accordingly, the basic strategy of this species is to rely on this excedent in available food.

Diet

The diet of *M. murinus* appears to be fairly similar to that of *C. medius.* Since the two species mainly feed on the same seasonal stock of "surplus" food, a more detailed analysis of food choices and of other behaviors related to feeding activity is necessary to explain the persistence of both species in this particular food niche.

Fruits form the staple diet of *M. murinus,* but are complemented with a higher proportion of animal matter than for *C. medius.* Chitin was present in all fecal samples of *M. murinus* trapped in the Marosalaza forest, and again Coleoptera were the most abundant insect food. Feces of at least one animal (which had no time to digest the bait used in the trap) were exclusively composed of chitin. Direct observations provided evidence of active predatory behavior directed toward rapid prey such as moths, crickets, and cockroaches. We also observed a chameleon devoured in small pieces during a period of 20 minutes. It is not surprising to find in the diet of *M. murinus* a relatively larger proportion of insects than in that of the larger *C. medius,* because, on the average, both species collect approximately the same absolute quantity of animal prey (cf. Hladik and Hladik, 1969; Charles-Dominique, 1977, for discussion on insect and fruit eater).

Together with the more rapid movements allowing capture of flying insects, the foraging habits of *M. murinus* (Fig. 7), exclusively observed in the dense undergrowth (at 1.5–6 m), tend to reduce interspecific competition between these two Cheirogaleinae, because they prey on partly different invertebrate populations.

Nectars and the exudates of plants and insects were fed upon in an opportunistic manner. Accordingly, the diet includes certain typical foods of *Cheirogaleus, Phaner*, and *M. coquereli* (see below). Green vegetal material

Fig. 7. *Microcebus murinus* foraging in the undergrowth of the Marosalaza forest.

was also occasionally included in *M. murinus* diet, as was previously observed by Martin (1973) in the south of Madagascar. A differential utilization of flower nectar and pollen may be due to plant morphology. For example, the Maronono, *Evonymus pleurostyloides* (43), a small tree, bears large bunches of flowers emanating from the trunk near the ground. These flowers were visited frequently in December/January during the different bouts of foraging activity always occurring at low level. The fruits of this tree, available in March, were eaten as well (G. F. Pariente, personal communication).

The seasonal variations in dietary composition of *M. murinus* appear to be roughly similar to those of *C. medius,* and depend on variations of food availability; but the increase of protein intake in spring and summer is more important.

Variations in Body Weight

The records of body weight of 72 recaptured *M. murinus* are presented in Fig. 8. Annual variations, although smaller than for *C. medius,* are still very important. Fat is accumulated under the skin and in the tail during the austral spring and summer. However, the tail fattening of *M. murinus* is never as spectacular as in the peculiar case of *Cheirogaleus*. Maximum variations recorded in the Marosalaza forest were from 2 to 9.5 cm^3. In captive conditions, *M. murinus* were fatter, and tail volume varied from 5 to 20 cm^3.

As in the preceding case study, the observations of captive animals during several years allowed a better understanding of the behavioral and physiological cycles. The period of increased food intake of *M. murinus* in artificial conditions occurred about 2 months after that of *C. medius* and was also preceded by a reduction of overall activity (Chapter 6). In these experimental conditions, fattening was followed by a type of lethargy which seems almost identical to that observed in the Marosalaza forest. Further observations on food intake (Chapter 8) also showed a variation in the propensity for different dietary components, according to the period of year.

In spring and summer, the diet of *M. murinus* included more protein as soon as insects were available in the Marosalaza forest. Similar variation in protein content of the overall diet was also observed during artificial food choices in standard conditions. The opposite dietary tendency (a diet rich in carbohydrates) appeared before the period of lethargy.

From December to May, the body weight variations resulting from these successive changes in feeding behavior of *M. murinus* (Fig. 8) are more

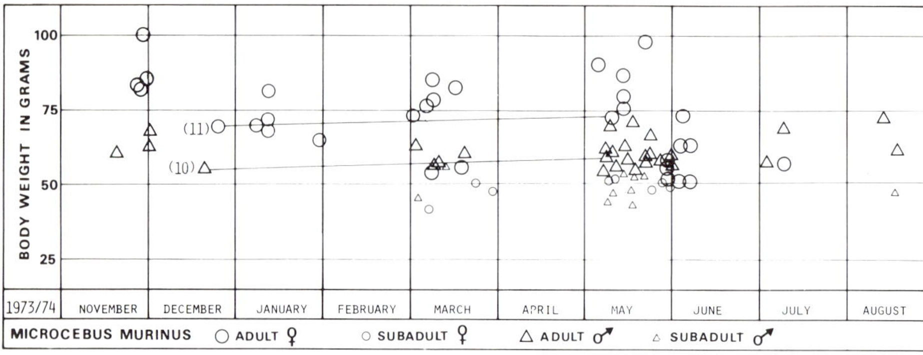

Fig. 8. Body weight of *Microcebus murinus* in the Marosalaza forest. The seasonal variation of body weight is smaller than that of *Cheirogaleus medius* (compare with Fig. 5). Individual variation is shown by solid line for one adult male (10) and one adult female (11).

progressive than for *C. medius*. The abrupt body weight increase following dietary variation, which was observed in autumn in artificial conditions only, probably also occurs in the wild and might explain the apparent scatter of the data collected in May (austral autumn) in the Marosalaza forest.

Protein intake from a standard diet was always larger for *M. murinus* (mean: 11.6%) than for *C. medius* (mean: 9.2%). This reflects the conditions in the Marosalaza forest where ecological mechanisms determine the possible diet for each species according to body weight and food availability. It also implies physiological and/or behavioral systems which are a sort of "preadaptation" directed towards the adequate food choices.

In addition to these variations in body weight, food intake, and food composition, which also depend on variations of the basal metabolism, Perret (1979) recorded a diminution in water intake during a period corresponding to the dry season. This appears as another kind of variation in physiological parameters which can be qualified as preadaptation to the environmental changing conditions.

Population Parameters

Population density of *M. murinus,* which approximates four animals per hectare (Table III) has the same order of magnitude as in other dry forests of the south of Madagascar (Charles-Dominique and Hladik, 1971). Thus, due to small body weight, the biomass of *M. murinus* is smaller than that of *Cheirogaleus medius*. For both species, the carrying capacity of the environment is likely to be determined by insect food availability, and the larger the insect component in the diet, the smaller the biomass.

According to our data on capture, recapture, and localized direct observation of marked animals, we only can suggest that the home range does not excede 150 m in diameter (about 2 ha) in the Marosalaza forest. In the more humid forest of the southeast of Madagascar where Martin (1973) carried out similar observations, the home range of *M. murinus* appeared to be smaller, possibly due to more important food availability. In these conditions, social structure involved "population nuclei" based on females, and the presence of "central males" with a relatively high body weight in spite of annual variation, and "peripheral males" of relatively lower body weight. A similar organization is expected to be found in Marosalaza.

Utilization of Insect Secretion: The Coquerel's Mouse Lemur, *Microcebus coquereli*

In contrast to precedent species of Cheirogaleinae, *Microcebus coquereli* shows little annual variation in food intake and body weight (body weight range: 280–335 gm in captured animals). Population density and biomass are consequently low as compared to those of other species (Table III), because the carrying capacity of the environment is determined by minimum food availability during the dry season. The species is adapted in an unusual manner to a peculiar food which remains available throughout the dry season: insect secretion. This was first described by Petter *et al.* (1971). A more detailed study of the population of *M. coquereli* of the Marosalaza forest is presented by Pages (Chapter 4). A few complementary points will be presented in order to stress the differences with sympatric prosimians in terms of feeding strategy.

Diet

According to the frequency of chitin in fecal samples, *M. coquereli* seems to be as insectivorous as *M. murinus,* and also eats large amounts of fruits (only one sample out of seven contained fruits only and no chitin). However, the liquid part of the diet cannot be measured or even detected in fecal samples. Only direct observation permitted an estimate of the importance of this food in terms of feeding frequency or time-budget. For instance, during the dry season *M. coquereli* was most frequently seen licking the secretion of the colonial aggregates of *Flatida coccinea* (up to 50% of the feeding observations of Pages, Chapter 4, and 60% of the feeding time). As a matter of fact, even in austral spring and summer, the Coquerel's Mouse Lemur ate many kinds of insect secretions (from cochineals and homopterans) as well as plant exudates, including gum of *Terminalia* spp. (48/49). For example the Fangoky, *Delonix boiviniana* (37), was frequented when flowering, the animals licking flowers (nectar) as well as leaves (cochineal exudate?).

Animal foods are of paramount importance because plant exudates and insect secretions have an extremely low protein content (Table I). The data of Pages (Chapter 4, Table I) also suggest that invertebrate food availability determines the distribution of *M. coquereli* in the study area. Vertebrate prey are also frequent, as shown by the skins of chameleons found in two out of seven feces samples. In captivity, *M. coquereli* eat young mice and any kind of mince meat. In fact, an unfortunate involuntary experiment at the Brunoy animal house showed that *M. murinus* can be preyed upon by *M. coquereli* if they are introduced into the same cage. In natural conditions, *M. coquereli* forages at relatively low levels (about 5 m in height), but it is likely that *M. murinus,* thanks to its small size and vivacity, can escape in the dense undergrowth or in small tree holes. As in most cases, predation might only concern a marginal part of the population (very young, or old animals).

Seasonal Cycles

As with other prosimian species, there are seasonal changes in the diet of *Microcebus coquereli* depending on food availability. The honey dew of *Flatida coccinea,* which is intensively consumed during the dry season, is exclusively secreted by larvae. In December/January, these Homoptera reach their imaginal stage (Fig. 3) and are no longer visited. At this time, the types of food also utilized by other prosimians (fruits and insects) are eaten. Accordingly, the diet of *M. coquereli* is subject to partly similar variations.

The variations recorded during the study of food intake in artificial conditions (Chapter 8), although of lesser importance than for other species, followed a significant pattern which evokes field conditions: a relative decrease of protein-rich foods in autumn and winter resulting from a diet mainly based on soluble carbohydrates, and the progressive increase of more nutritious foods (protein and fat) in spring. As for other species, this seasonal variation of the "dietary tendency," which was maintained even in standard and homogeneous conditions, appears as an adaptation to the extremely important variations in composition of the food available in the forest of the West Coast of Madagascar.

Utilization of Gums: The Fork-Marked Lemur, *Phaner furcifer*

Phaner furcifer has approximately the same body size as *Microcebus coquereli,* and is also able to face the dry season without apparent variation in body weight and no evident change of the activity pattern and/or basal metabolism. Nevertheless, there is no significant competition during the dry season for the staple food of each species since *P. furcifer* is highly specialized on gum and other tree exudates, as was shown by Petter *et al.* (1971), and more recently observed in the Marosalaza forest by Charles-Dominique and Petter (Chapter 3). Different plant exudates and insect secretions can actually be eaten by both *P. furcifer* and *M. coquereli,* but behavioral and/or physiological specialization result in a definite interspecific difference in dietary preference.

As in the preceding case, the staple food which is relatively poor in protein, has to be complemented with animal food (mainly insects). Every night, *P. furcifer* forages for insects during long periods but always subsequent to an initial period of gum feeding (Chapter 3). The selection of prey is different than that of *Microcebus coquereli;* in particular, no vertebrates are eaten. Foraging also occurs at a higher level (8–10 m, and often in the tree tops). Interspecific competition is thus reduced; but, as for *M. coquereli,* the carrying capacity of the environment is probably determined by insect food availability. The relatively important biomass of *P. furcifer* in the Marosalaza forest (about twice as large as that of *M. coquereli;* Table III), may result from the relative protein richness of the gum of *Terminalia mantaliopsis* (48) as compared to insect exudate (Table I). This gum necessitates a smaller proportion of insects (about 10%) in the diet of *P. furcifer.*

In captivity (see Chapter 8), the artificial diet was probably too different from that in nature to allow a normal physiological cycle. No obvious seasonal pattern was observed.

Phaner furcifer has no imperative necessity to find hollow trunks for protection during the dry season (as for *Cheirogaleus medius,* see above). Nevertheless, the number of adequate hollow trunks to use as day nests may be limited, due to occupation by a large number of *Lepilemur ruficaudatus.* As a matter of fact, *P. furcifer* was so frequently found in the abandoned nests constructed by *M. coquereli* during the first survey of Petter *et al.* (1971) that it was not possible to distinguish which species was responsible for nest construction.

Folivory: The Sportive Lemur, *Lepilemur ruficaudatus*

The very high biomass of *Lepilemur ruficaudatus* (body weight about 900 gm, Fig. 9) in the Marosalaza forest (Table III) is obviously related to the ability of the animal to utilize the most ubiquitous food resource: leaves.

In the fecal samples collected during the austral summer, leaf fiber was mixed with the seeds of fruits, especially those of *Diospyros* spp. (55/56), but leaves always constituted the staple food. During the austral winter, the limited number of evergreen species, and the frequency and high amounts of alkaloids present (see Chapter 1) may reduce the possible food choices of *Lepilemur ruficaudatus.* In practice, leaves and/or flowers of one or two species may be sufficient to supply the Sportive Lemur population during

Fig. 9. *Lepilemur ruficaudatus* in the Marosalaza forest (photo by C. M. Hladik 1201 P).

several dry months, as was demonstrated for *L. leucopus* (Charles-Dominique and Hladik, 1971).

In all species of *Lepilemur* thus far studied, caecotrophy is observed (Hladik and Charles-Dominique, 1974; C. M. Hladik, 1978a), which allows sufficient efficiency in digestion of long-chain β-linked carbohydrates. *Lepilemur ruficaudatus* was also observed in the animal house of Brunoy reingesting "feces" at various intervals, during the quiet period of artificial daylight.

The survey of food intake in artificial conditions (Chapter 8) showed a definite seasonality. As in the case of the other nocturnal prosimians subject to important seasonal changes, there was an increase of the proportion of fruit included in the diet during spring and summer. Although the annual variation of body weight and activity is not very marked (Chapter 6), *L. ruficaudatus* appears to be "preadapted" through a seasonal shift in feeding behavior, to include a larger proportion of fruits in its diet at the time of maximum availability.

The reproductive cycle (Chapter 6), which differs from that of the sympatric species, also permits an adequate utilization of food availability: females lactate at the time of leaf growth (young leaves have maximum protein content), and the young are weaned when a large surplus of food is available to all species.

DISCUSSION

The feeding strategies of the five nocturnal prosimian species inhabiting the dry deciduous forest of Marosalaza, near Morondava, can be illustrated by the average annual food intake or "dietogram." In Fig. 10, the dietograms proposed for the Malagasy species are only approximations, since the present study did not allow exhaustive measures of food intake. These estimates are compared to the results of Charles-Dominique (1977) concerning the diets of five other nocturnal prosimians (*Perodicticus potto, Euoticus elegantulus, Galago alleni, Galago demidovii,* and *Arctocebus calabarensis*) inhabiting the Gabon rain forest, where collection of stomach samples allowed more accurate measures of the quantity of the different food categories actually consumed.

Diets of the different species are related to the logarithm of the biomass (resulting in the definition of dietary grades, see above), because the carrying capacity of the environment depends on the overall production of leaves, fruits, and insects. Even in the two contrasting biotopes compared in Fig. 10, species of similar diet have a biomass of similar order of magnitude; for example, *Euoticus elegantulus* and *Phaner furcifer* both

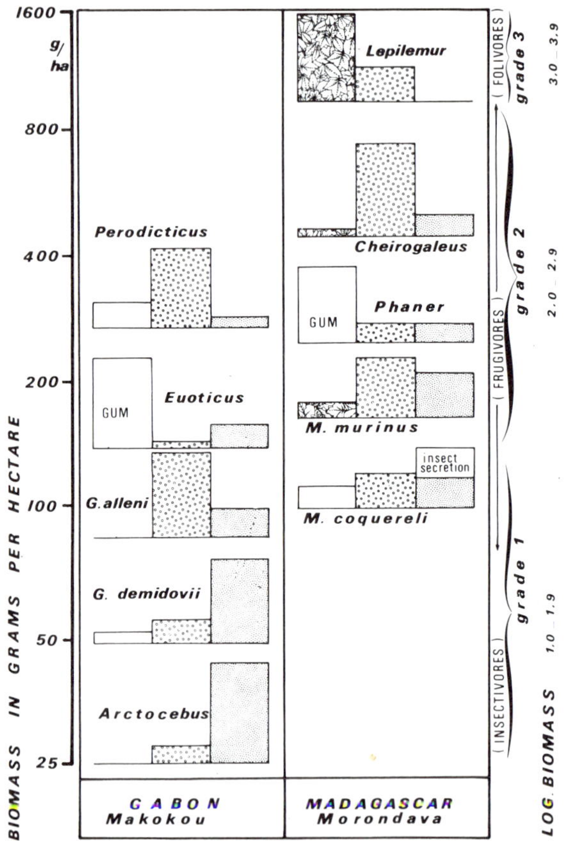

Fig. 10. A comparison of diets of the nocturnal prosimian species of Madagascar (Marosalaza forest) to those of the Gabon rain forest (after Charles-Dominique, 1977). Each species' diet is illustrated by the relative amount of leaves and/or gums (left rectangle), fruits (central rectangle), and animal matter (right rectangle) which have been ingested.

specialized on gum supplemented with invertebrate food (Chapter 3). Minor differences may result from the larger number of sympatric species inhabiting the rain forest, each of which has a more specialized diet and thus a more restricted quantity of potential food resources (Hladik and Chivers, 1978).

A characteristic pattern of food choices allows each species to be maintained in this complex and dynamic pyramid of producers and consumers. The mechanism allowing a certain flexibility together with a constant overall feeding tendency may involve different intensities in the respective effects of immediate hedonistic conditioning by taste stimulation, and long-term conditioning from food digestion and absorption (C. M. Hladik,

1978b). For instance, *Lepilemur ruficaudatus* is certainly mostly motivated by long-term conditioning after digestion of leaf material, whereas for *Microcebus* spp. the motivation to search actively for more nutritious fruits and insects may be related to a higher degree of taste stimulation and subsequent conditioning. In fact, this very crude analysis of feeding behavior does not seem sufficient to explain subtle differences nor constant overall tendencies in species such as *Phaner furcifer* and *Microcebus coquereli* which feed on many similar foods, but have nevertheless definite species preferences.

Seasonal Variations of the Dietary Tendencies

The dietary tendencies manifested by the different prosimian species present annual variations which have been qualified as preadaptations because, even in artifical conditions, the patterns of food choice vary according to the expected variations in food availability in the natural environment (Chapter 8). Such regular variations in feeding behavior may also exist in other species subject to regular and important annual changes of the environment. Nevertheless, the very marked variations in food availability and food consumption in the forest of the West Coast of Madagascar probably exert an intensive selective pressure which explains why the characteristic "seasonally variable feeding behavior" is firmly established in the five prosimian species thus far studied. However, the relative contributions of genetic inheritance and experience in the development of these patterns cannot yet be ascertained.

Further behavioral characteristics related to utilization of space and time reduce interspecific competition among the nocturnal prosimians of the Marosalaza forest by completing the overall differences in feeding behavior.

Utilization of Different Heights for Foraging

The different patterns of foraging described in this paper are illustrated in Fig. 11 by the mean heights and mean diameter of the supports on which the prosimians have been observed in the forest. Significant differences were determined by the t test applied to all possible pairs of species (calculation of t according to the means and standard deviations as in Chapter 8).

The differences in height are not significant for the species which have a feeding strategy sufficiently different to avoid competition; for instance, *C. medius* and *M. coquereli* ($p > 0.1$) which feed simultaneously only during the period of maximum (or surplus) food availability. In contrast, highly

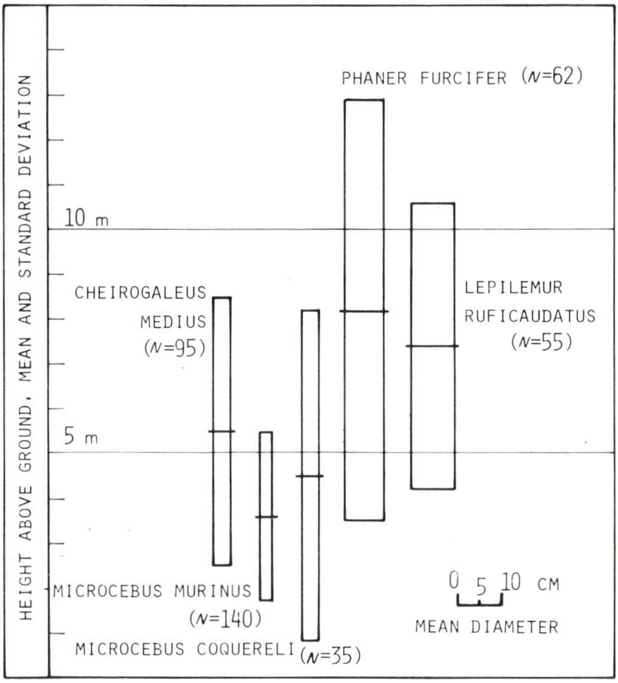

Fig. 11. Field location of the different nocturnal prosimian species in the Marosalaza forest. The height above ground where the animals were first observed are compared within species, and the thickness of each column is proportional to the diameter of the support used by the animals (after cumulative data of P. Charles-Dominique and C. M. Hladik).

significant differences ($p<0.001$) were found between *M. coquereli* and *Phaner furcifer* which are likely to compete for invertebrate food during the dry season, and between *C. medius* and *M. murinus* which have very similar food choices and similar yearly cycles.

The utilization of supports of different diameter merely reflects forest structure, with a significant difference ($p<0.01$) between *M. coquereli* and *P. furcifer* which also feed at very different heights.

Circadian Rhythms Considered in Terms of Food Niche Separation

The definite patterns of overall and feeding activity found by Pages and Petter-Rousseaux (Chapter 7) in the five species of nocturnal prosimians, also appear to be complementary to other distinctive adaptations reducing food competition.

The differential distribution of locomotor and feeding activity throughout the night tend to separate *C. medius* (mainly feeding at the end of the night) and the two *Microcebus* species (maximum activity at the beginning of the night, at least during periods of low food availability). By contrast, in *Phaner furcifer* the occurrence of gum feeding at the beginning of the night appears to be related mainly to intraspecific competition since individuals share (to a certain extent) the same trees producing gums (Chapter 3).

Although most cases of niche separation due to utilization of food resources at different times are best documented for nocturnal versus diurnal species (Charles-Dominique, 1975), intermediate cases were observed between diurnal primate species with different activity rhythms for feeding (Hladik and Chivers, 1978). The present study suggests that circadian rhythms of nocturnal species may also constitute an important component of the feeding strategy.

Finally, in the nocturnal world where the different prosimians apparently have very similar visual efficiencies (Chapter 5), the visual learning capacities may be important for developing different feeding strategies, since interspecies differences exist in the utilization of the "integrated" visual information, reflected by variations in transfer of learning, memory, and forgetting (Chapter 10). Thus sensory processes involved in the search for food sources must be considered as part of the overall species strategy.

REFERENCES

Charles-Dominique, P. (1975). Nocturnality and diurnality: An ecological interpretation of these two modes of life by analysis of the higher vertebrate fauna in tropical forest ecosystems. *In* "Phylogeny of the Primates"(W. P. Lucket and F. S. Szalay, eds.), pp. 69–88. Plenum, New York.

Charles-Dominique, P. (1977). "Ecology and Behaviour of Nocturnal Primates." Duckworth, London.

Charles-Dominique, P., and Hladik, C. M. (1971). Le Lépilemur du Sud de Madagascar: Ecologie, alimentation et vie sociale. *Terre Vie* **25**, 3–66.

Debray, M., Jacquemin, H., and Razafindrambao, R. (1971). Contribution à l'inventaire des plantes médicinales de Madagascar. *Trav. Doc. ORSTOM.* No. 8.

Decary, R. (1946). Plantes et Animaux utiles de Madagascar. *Ann. Musée Col., Marseille* **54**, 1–234.

Eisenberg, J. F., Muckenhirn, N. A., and Rudran, R. (1972). The relation between ecology and social structure in Primates. *Science* **176**, 863–874.

Hladik, A. (1978). Phenology of leaf production in a rain forest of Gabon: Distribution and composition of food for folivores. *In* "The Ecology of Arboreal Folivores" (G. G. Montgomery, ed.), pp. 51–71. Smithson. Inst. Press, Washington, D.C.

Hladik, A., and Hladik, C. M. (1969). Rapports trophiques entre végétation et Primates dans la forêt de Barro Colorado (Panama). *Terre Vie* **23**, 25–117.

Hladik, A., and Hladik, C. M. (1977). Signification écologique des teneurs en alcaloïdes des végétaux de la forêt dense: Résultats des tests préliminaires effectués au Gabon. *Terre Vie* **31**, 515–555.
Hladik, C. M. (1975). Ecology, diet and social patterning in Old and New World Primates. *In* "Socioecology and Psychology of Primates" (R. H. Tuttle, ed.), pp. 3–35. Mouton, The Hague.
Hladik, C. M. (1977a). A comparative study of the feeding strategies of two sympatric species of leaf monkeys: *Presbytis senex* and *P. entellus*. *In* "Primate Ecology" (T. H. Clutton-Brock, ed.), pp. 323–353. Academic Press, New York.
Hladik, C. M. (1977b). Chimpanzees of Gabon and Chimpanzees of Gombe: Some comparative data on the diet. *In* "Primate Ecology" (T. H. Clutton-Brock, ed.), pp. 481–501. Academic Press, New York.
Hladik, C. M. (1977c). Field methods for processing food samples. *In* "Primate Ecology" (T. H. Clutton-Brock, ed.), pp. 595–601. Academic Press, New York.
Hladik, C. M. (1978a). Adaptative strategies of Primates in relation to leaf eating. *In* "The Ecology of Arboreal Folivores" (G. G. Montgomery, ed.), pp. 373–395. Smithson. Inst. Press, Washington, D. C.
Hladik, C. M. (1978b). Diet and ecology of Prosimians. *In* "The Study of Prosimian Behavior" (G. A. Doyle and R. D. Martin, eds.), pp. 307–357. Academic Press, New York.
Hladik, C. M. (1980). Diet and evolution of feeding strategies among forest primates. *In* "Human Evolution through Hunting and Gathering" (G. Teleki and R. S. O. Harding, eds.), Columbia Univ. Press, New York (in press).
Hladik, C. M., and Charles-Dominique, P. (1974). The behaviour and ecology of the sportive lemur *(Lepilemur mustelinus)* in relation to its dietary peculiarities. *In* "Prosimian Biology" (G. A. Doyle, R. D. Martin, and A. C. Walker, eds.), pp. 23–37. Duckworth, London.
Hladik, C. M., and Chivers, D. J. (1978). Ecological factors and specific behavioural patterns determining Primate diet (concluding discussion). *In* "Recent Advances in Primatology" (D. J. Chivers and J. Herbert, eds.), Vol. 1, pp. 433–444. Academic Press, New York.
Hladik, C. M., and Hladik, A. (1972). Disponibilités alimentaires et domaines vitaux des primates à Ceylan. *Terre Vie* **26**, 149–215.
Hladik, C. M., Hladik, A., Bousset, T., Valdebouze, P., Viroben, G., and Delort-Laval, J. (1971). Le régime alimentaire des Primates de l'île de Barro Colorado (Panama): Résultats des analyses quantitatives. *Folia Primatol.* **16**, 85–122.
Jolly, A. (1966). "Lemur Behaviour. A Madagascar Field Study." Univ. of Chicago Press, Chicago, Illinois.
MacArthur, R. H. (1965). Patterns of species diversity. *Bio. Rev.Cambridge Philos. Soc. Bio. Rev.* **40**, 510–533.
MacArthur, R. H., and Wilson, E. O. (1967). "The Theory of Island Biogeography." Princeton Univ. Press, Princeton, New Jersey.
Martin, R. D. (1973). A review of the behaviour and ecology of the Lesser Mouse Lemur (*Microcebus murinus* J. F. Miller 1777). *In* "Comparative Ecology and Behaviour of Primates" (R. P. Michael and J. H. Crook, eds.), pp. 1–68. Academic Press, New York.
Odum, E. P., and Odum, H. T. (1959). "Fundamentals of Ecology." Saunders, Philadelphia, Pennsylvania.
Perret, M. (1979). Seasonal and social determinants of urinary catecholamines in the Lesser Mouse Lemur (*Microcebus murinus,* Cheirogaleinae, Primates). *Comp. Biochem. Physiol.* **62**, 51–60.

Petter, J. J. (1978). Ecological and physiological adaptations of five sympatric nocturnal lemurs to seasonal variations in food production. *In* "Recent Advances in Primatology" (D. J. Chivers and J. Herbert, eds.), Vol. 1, pp. 211–223. Academic Press, New York.

Petter, J. J., Schilling, A., and Pariente, G. (1971). Observations éco-éthologiques sur deux lémuriens malgaches nocturnes: *Phaner furcifer* et *Microcebus coquereli*. *Terre Vie* **13,** 287–327.

Richard, A. F. (1974). Intraspecific variation in the social organization and ecology of *Propithecus verrauxi*. *Folia Primatol.* **22,** 178–207.

Sussman, R. W. (1974). Ecological distinctions in sympatric species of *Lemur*. *In* "Prosimian Biology" (R. D. Martin, G. A. Doyle, and A. C. Walker, eds.), pp. 75–108. Duckworth, London.

Sussman, R. W. (1978). Nectar-feeding by *Lemur mongoz mongoz* and its evolutionary and ecological implications. *In* "Recent Advances in Primatology" (D. J. Chivers and K. A. Joysey, eds.) Vol. 3. Academic Press, New York.

Whittaker, R. H., and Feeny, P. P. (1971). Allelochemics: Chemical interactions between species. *Science* **171,** 757–770.

Wilson, E. O. (1975). "Sociobiology. The New Synthesis." Harvard Univ. Press (Belknap), Cambridge, Massachusetts.

3 *Ecology and Social Life of Phaner furcifer*

P. CHARLES-DOMINIQUE and J.J. PETTER

INTRODUCTION

Phaner furcifer, a medium sized nocturnal prosimian of about 300 gm, is characterized by a black dorsal stripe which bifurcates on the crown into two stripes continuing to the eyes (= Fork-marked dwarf Lemur, Fig. 1). Originally classified in the subfamily Cheirogaleinae, *Phaner furcifer* is now considered as the only member of a distinct subfamily: Phanerinae (Petter *et al.,* 1977). The population inhabits different types of forests on the Northeast, North, and West Coasts of Madagascar (Petter *et al.,* 1977). In the West Coast *Phaner furcifer* is present in the deciduous forest as well as in the transitional zone between coastal lagoons and forest (see Chapter 1).

The present study was undertaken in the Marosalaza forest in the private reserve of Mr. Dehaulme at Beroboka (Analabe), 50 km north of Morondava. A field study on the ecology of *Phaner furcifer* and *Microcebus coquereli* was conducted 3 years previously in the same area, between November and December, 1970 (Petter *et al.,* 1971).

The present work is based on several stays at Beroboka by the authors (J.J. Petter: October 15–November 10, 1973 and January–February, 1974; P. Charles-Dominique: November 2–December 2, 1973 and May 1–June 15, 1974). In addition, supplementary observations in Beroboka by our colleagues C.M. Hladik, G.F. Pariente, E. Pages, and R. Albignac on diet and vocal behavior complete our data, thus covering practically the complete annual cycle from October 1973 to August 1974.

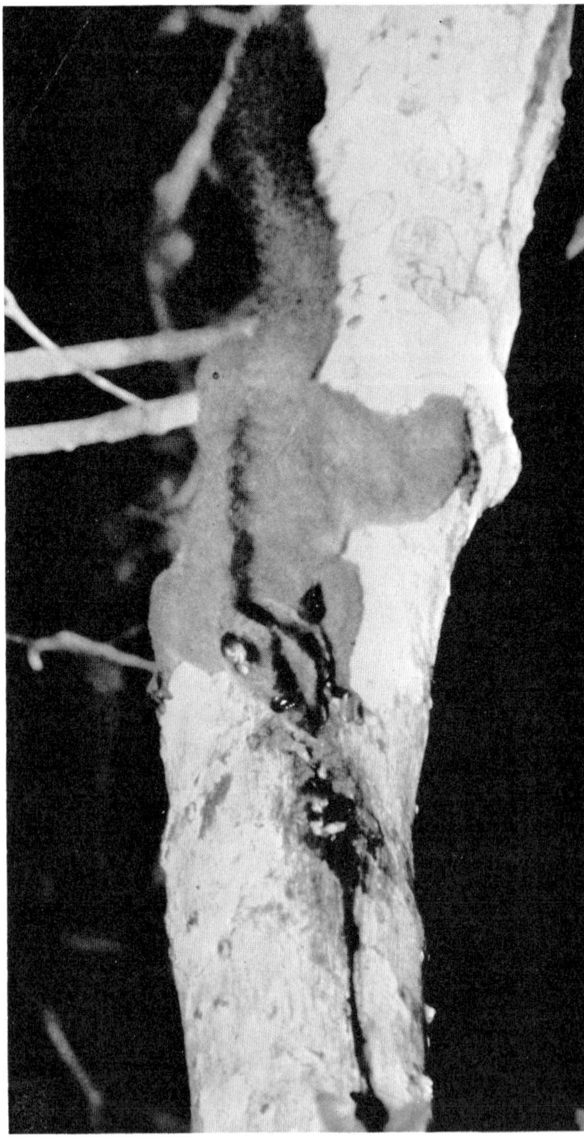

Fig. 1. *Phaner furcifer* feeding at night on gum of a Talinala tree, *Terminalia mantaliopsis* (No. 48 in Appendix I, Chapter 1) (photo by C.M. Hladik).

METHODS

The study area, described by A. Hladik (Chapter 1), was divided by 8080 m of narrow trails, cut in a rectilinear pattern, with intersections every 40 or 80 m. Plastic markers were placed on the trails at intervals of 20 m and a map of the area was drawn with localization of trails and of important trees. The study area of *Phaner furcifer* was about 500 × 500 m.

The first observations were made by following the animals at night with the aid of a head lamp. This method permits observations on the diet, locomotion, vocal behavior, and activity, but individual identification is generally difficult. In May–June, 1974, we began a technique of capture and marking individual animals: several nooses attached to a string were placed on the branches of *Terminalia* spp. trees, near the places of gum production. *Phaner furcifer* regularly visits these trees, principally at the beginning of the night; the captures were made by pulling the string when an animal passed through the noose. Different trees were equipped to permit a survey of all the individuals present in the study area. Immediately after capture the general morphology of the animal was examined (weight, teeth, genital organs, mammary glands, and scars), feces were collected for further examination, ears were marked by notches, and the tail by shaving the fur according to various patterns. Nine individuals were marked, and five others could be identified by scars.

Like many other Malagasy prosimians, *P. furcifer* is not shy and in the western dry forest visibility conditions are quite good. In addition, all movements of *P. furcifer* are accompanied by a weak vocalization ("Hon") which permitted, with some practice, an identified individual to be followed during long periods and the itineraries to be plotted on a map.

DIET

Phaner furcifer is a typical gum eater, and, in the Marosalaza forest activity is devoted to visits of *Terminalia* spp. trees (Talinala in Malagasy language pronounced Taly; cf. No. 47 and 48 in Appendix I, Chapter 1). These small trees of the Combretaceae family are generally parasitized by a Coleopteran larvae living between the wood and the bark. The worms burrow galleries beginning in a rectilinear pattern (15–20 cm) and finishing in a spiral of about 3 cm in diameter. The galleries are visible on the bark as two parallel dotted lines spaced 2 mm apart, each dotted line being composed of tiny orifices exuding the gum which accumulates in amber-colored drops. Some *Terminalia* trees also show large sores which produce gums.

In the fall (March–May), gums provide the bulk of the vegetal food intake of *Phaner furcifer* (202 observations) and it is noteworthy that a bird, *Coua cristata,* exploits this same resource during the daytime period (Charles-Dominique, 1976; Fig. 2). Certain other trees are also exploited for gums: *Commiphora* sp. (Burseraceae No. 25 = Harofymena, four observations), *Colvillea racemosa* (Caesalpiniaceae No. 35 = Sarongaza, four observations); for bud exudation: *Zanthoxylum tsihanimposa* (Rutaceae No. 154, Monongo, one observation); or for the sap: *Adansonia* spp. (Bombacaceae No. 20–21 = Reniala, 23 observations). In November, 1970 (austral spring), in addition to *Terminalia* spp. gum which provided the bulk of diet, *Phaner furcifer* was observed by Petter *et al.* (1971) licking flowers of *Crateva greveana* (Capparidaceae = Tambitsiky) and "syrup" produced by an insect larvae of the Machaerotidae family (Homoptera, Auchenorhynche). These larvae live in small lodges grouped on thin branches of *Rhopalocarpus lucidus* trees (Rhopalocarpaceae = Talafoty). *Phaner furcifer* licked the syrup exuding from the lodges without ingesting the insects. In the 1973–1974 study area, 3 km farther to the east, Tambitsiky and Talafoty trees were not present.

Considering the various types of forest inhabited by *Phaner furcifer* (i.e., dry forest on the West Coast, rain forest on the East Coast), this species necessarily feeds upon different tree species, according to the location. However, it is likely that gums and vegetal exudations compose the main part of the diet. Compared to other prosimians of the same size, *P. furcifer* has a long "tooth comb" (Petter *et al.,* 1971); this peculiarity discussed by Martin (1972b) and Charles-Dominique (1977a) is an adaptation for gum collection, and is also present in *Galago (Euoticus) elegantulus,* a highly specialized gum eater among African Lorisidae.

Gums are usually eaten when they seep directly on the bark or exude from crevices (Fig. 1), but often insect galleries are actively opened. In the latter cases, *Phaner furcifer* uses its long tongue and also the narrow and procumbant tooth comb to scoop the gum which would otherwise be inaccessible without such a "tool." *Phaner furcifer* is also characterized by the development of the upper first premolar (caniniform). This peculiarity, seen only in *Allocebus trichotis,* among Malagasy lemurs and *Galago (Euoticus) elegantulus,* among African lorisids, appears to be an adaptation for the extraction of vegetal exudations. In addition to this specialized dentition, *P. furcifer* has long sharpened nails which permit access to smooth trunks and large branches where gums are located.

At the beginning of the twilight, about 15–20 minutes before complete darkness, *Phaner furcifer* leaves the den, emits numerous calls and then visits *Terminalia* spp. trees producing gums. In the 25-ha study area, only 43 of the 69 Taly trees present actually produced gums. Thus, for the 14

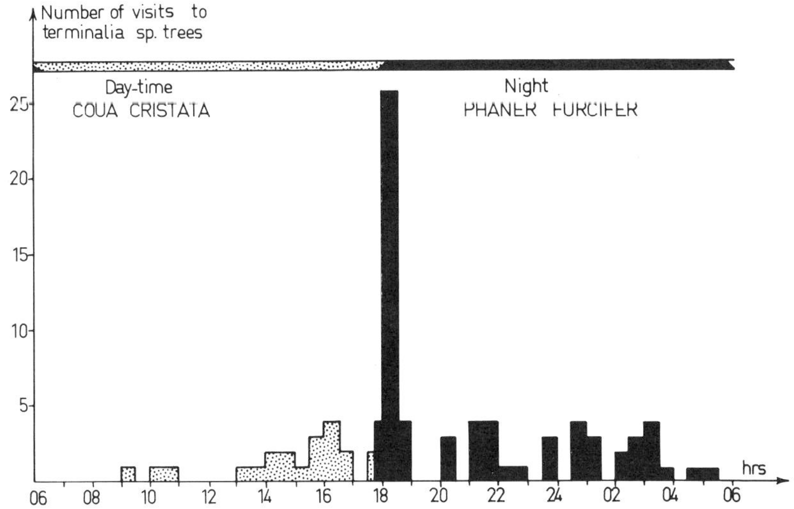

Fig. 2. Frequency of visits to gum-producing *Terminalia* trees, during the day by a bird, *Coua cristata* (shaded) and during the night by the prosimian *Phaner furcifer* (darkened).

animals inhabiting this area there is an average of 3 gum-producing Talys per individual. However due to the overlapping of male and female territories (Fig. 5 and 6) each individual has access to 9 gum-producing Talys shared between males and females. The quantity of gum produced by each Taly is very variable. *Phaner furcifer* seem to know perfectly well the localization of gum-producing trees and the better points of exudation and gum accumulation, as evidenced by the fact that they rapidly visit these trees in priority following regular itineraries. Later, they return to these trees to collect some drops of gum left after the first visit and also exploit other tree species producing gums or vegetal exudates. The learning and memorization of the complex itineraries appear to depend on vision, and corresponds to the type of visual learning capacity demonstrated by this species (Chapter 10).

Several other primate species have also been observed to eat, occasionally or regularly, gums of only a limited number of tree species among a large diversity of vegetal exudates (gums, resins, and sap): Mimosaceae gums *Galago senegalensis* (Sauer and Sauer, 1963); *Galago (Euoticus) elegantulus* and *Perodicticus potto* (Charles-Dominique, 1971, 1977a); *Galago crassicaudatus* (Bearder and Doyle, 1974); *Papio papio* (Hausfater and Bearce, 1977); Anacardiaceae, Combretaceae and Palmaceae gums *Callithrix jachus* (Coimbra-Filho, 1972); *Cebuella pygmaea,* (Moynihan, 1976).

Preliminary analysis of some gums eaten by primates indicates that gum chemistry is very complex. These hydrosoluble substances are mainly composed of highly polymerized C_6 and C_5 sugars: galactose, rhamnose, arabinose (work in progress, and Coimbra-Filho, 1972), and should not be confused with the nonhydrosoluble resins composed of terpenes. Gums are extremely poor in nitrogen, and *Phaner furcifer,* as well as other gums eaters, normally cover their protein requirement with animal prey. However, contrary to *Microcebus coquereli, P. furcifer* seems very selective in prey choice: two captive animals would not pay attention to large Coleopteran larvae, grasshoppers, small reptiles (Geconidae and Cameleonidae). Nevertheless, they immediately ate a large Sphingid moth and a praying mantis. According to the fact that this species begins its nocturnal activity by visiting gum trees, hunting behavior is often unnoticed. It is generally after 22.00 hours that *P. furcifer* can be observed hunting prey in the foliage or directly on tree trunks. Capture is achieved by a rapid movement of the hands, identical to the typical Cheirogaleinae and Galaginae pattern, after which the prey is transferred to the mouth. All *P. furcifer* feces collected in the wild contained fragments of insect chitin.

POPULATION PARAMETERS

Classical techniques of density evaluation (Charles-Dominique, 1971, 1977a) are different to apply to *Phaner furcifer.* The animals are very active during the night and often, excited by the light beam, come toward the observer and alarm call vigorously. Calculation of population density by counting all individuals located along a forest trail explored with a head lamp is thus biased by this behavior and may lead to an overevaluation. Sometimes *P. furcifer* seems to be more abundant in small areas where numerous gum trees are found, and observations made in 1970 (Petter *et al.,* 1971) were probably in such conditions. In this case, the population densities were calculated after extrapolation of data obtained in areas too small in size (1 ha and 2 ha) and thus cannot be considered as representative of a mean population density. Our study area of 25 ha contained 14 *P. furcifer* of which all were individually identified. Extrapolation gives a value of about 50–60 animals/ km^2, a population density higher than that of the African lorisids gum eater *Galago (Euoticus) elegantulus* (15–20/km^2, Charles-Dominique, 1977a).

Phaner furcifer seems to have a typical breeding season, identical to those of other Malagasy prosimians which breed during austral spring (Petter-Rousseaux, 1962, 1968, 1974). In October–November, 1973, no young were observed; in May, 1974, four of the five adult females observed in the study area were accompanied by one juvenile of about 5 months of age. Later, in July 1974, a juvenile of about 2 months was caught in an adjacent area (probably a late birth as often occurs in other Malagasy prosimians).

Little information is available concerning predation. However, one of the authors (J.J. Petter) has observed a bird of prey (*Aviceda madagascariensis*) catching a *P. furcifer* at the beginning of twilight. After a short fight in the air, the *P. furcifer* succeeded in escaping and fell from a height of about 20 m into the vegetation. He was probably wounded and remained for some time on a branch, licking his flank. A diseased or young animal would probably have been killed in such conditions. It is interesting because two nights previously this individual was noticed to have begun activity rather early in the evening. This example of predation indicates a selective aspect of the environment limiting the animal to nocturnal activity. During nocturnal activity, the large owl *Asio madagascariensis* could be a potential predator, as well as the Boidea snake *Sanzinia madagascariensis*.

Phaner furcifer is a typical nocturnal prosimian in which the activity rhythm depends upon variations in luminosity (Pariente, 1974, and Chapter 5). When the luminosity falls below 8 lux (about 15–20 minutes before full night-time), the animals leave their tree holes. After an initial peak of activity, at the beginning of the night, activity continues at a constant level until early dawn when they return to their den. During the daytime period, *P. furcifer* sleeps in tree holes generally situated in large trees (Harofy, Baobab) or sometimes an abandoned nest of *Microcebus coquereli* is used (Petter *et al.*, 1971). In the breeding colony at Brunoy, *P. furcifer* rarely carry leaf material into their boxes as do many other prosimian species (*Microcebus, Galago*).

SOCIAL LIFE

Social Communication

Vocal Signals

Vocalizations are predominant in *Phaner* social communications: a mean of 30 loud calls or series of loud calls/hour has been counted in a radius of about 200 m in the Marosalaza forest.

"Hon" call: The male and the female of a pair are in constant vocal contact, often the male following the female. All movements are accompanied by a weak "Hon" (Fig. 3A) similar to the contact call of numerous gregarious animals (i.e., *Lemur* sp., Simian species). Movements are

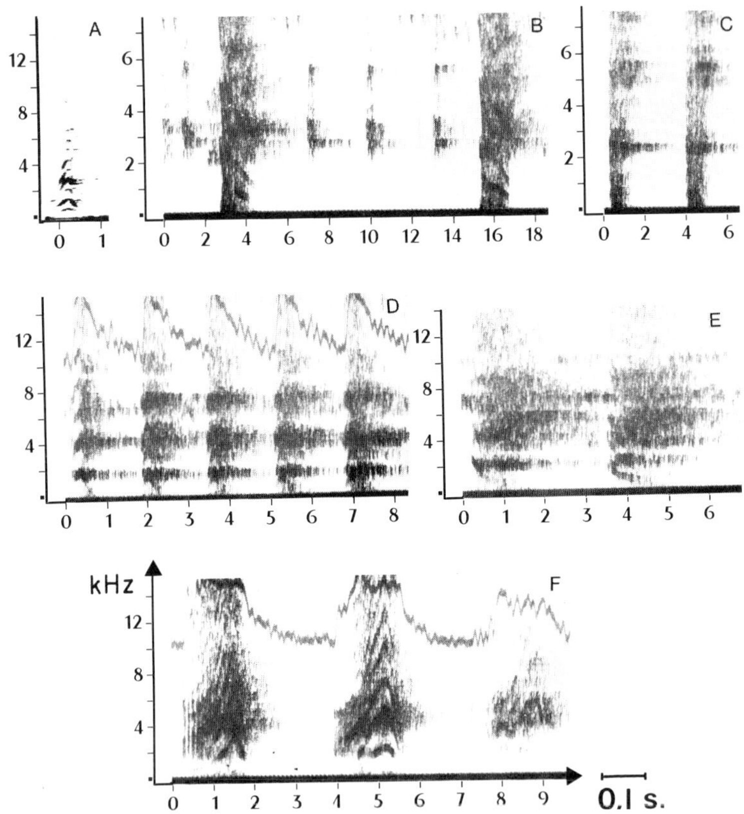

Fig. 3. Vocalizations of *Phaner furcifer* as shown on sonograms in KHz (vertical axis) according to time (horizontal axis in 1/10 of second). (A) The weak vocalization "Hon" accompanying movements. (B) The "Ki" call is shown as the more frequently repeated signal superimposed with a loud wide frequency signal ("Kea" call) emitted simultaneously by another animal. (C) Transitional signals between "Ki" and "Kiu" calls. (D) Series of "Kiu" calls. (E) "Kea" male loud call emitted twice in succession. (F) Distress calls emitted in a situation of fear. The continuous line in D and F is a complementary measure of energy distribution.

automatically accompanied by the weak "Hon," even if an animal (male or female) is separated by 50–60 m from the mate.

"Ki" and "Kiu" calls: The "Hon" can increase in intensity, evolving to a kind of "Ki," yielding a sonogram composed of a parabola and several overtones (energy mainly distributed between 500 and 5000 Hz). Rapidly, the "Ki" evolves toward a loud guttural "Kiu" repeated more or less rapidly in series of 5–25 units (Fig. 3C and D), sometimes more. At a low level of excitation, the "Kiu" is composed of an acute parabola with

overtones but at a high level of excitation the structure is more complex. When a male or a female emits a series of "Kiu" calls the partner answers systematically with the same calls, even if they are separated by 50–60 m, which may be continued for 10 seconds to 1 minute in a duet. Often these calls elicit other similar calls and duets in the vicinity. These calls which reflect the state of excitation of the emitters probably give information as to the identity and position of the partner, as well as the neighbors (Fig. 9 and 10). The distribution of "Kiu" calls varies during the activity period (Fig. 8): a short peak following awakening, a minimum between 19.00 and 21.00 hours (feeding activity) and an increase in the middle of the night (42% between 21.00 and 00.00 hours corresponding to the "meeting" in overlapping zones; see below).

"Kea" or male call: This loud call is exclusively emitted by males, generally as a single unit, sometimes repeated two or three times in succession (Fig. 3E). The distribution of males calls during the night follows the same pattern as "Kiu" calls, with a higher frequency at awakening (Fig. 8). The male call elicits other male calls in the vicinity, as well as "Kiu" series. This call can be considered as being derived from "Ki" and "Kiu" calls, with a reinforcement of the descending part of the parabola.

"Bleating" call: A female associated with a juvenile infant emits, from time to time, a loud vocalization roughly recalling the bleating of a goat.

Fighting call: Like other prosimians, *Phaner furcifer* has a typical vocalization associated with fighting and defense. This vocalization is composed of a bimodal call corresponding to expiration and inspiration: "Hein-ein-Hein-ein-Hein-ein."

Distress call: In situations of pain or great fear (handled animal), *Phaner furcifer* can emit a typical distress call (Fig. 3F), a kind of ascending tremolo moan similar, though of lower pitch, to the calls emitted by other nocturnal prosimians in such situations (Petter and Charles-Dominique, 1978).

Olfactory signals

Contrary to almost all other nocturnal lemurs (Lorisidae, Cheirogaleidae) urine and feces seem to play no role in *Phaner furcifer* social communications: no behavior such as urine washing, rhythmic micturition, anogenital dragging or other patterns of marking were observed in association with urination or defecation (Schilling, 1978). Urine and feces fall directly to the ground, and no particular sites of defecation or urination have been noticed.

Phaner furcifer has a large cutaneous throat gland, which is well developed in the male and atrophied in the female (Rumpler and Andria-

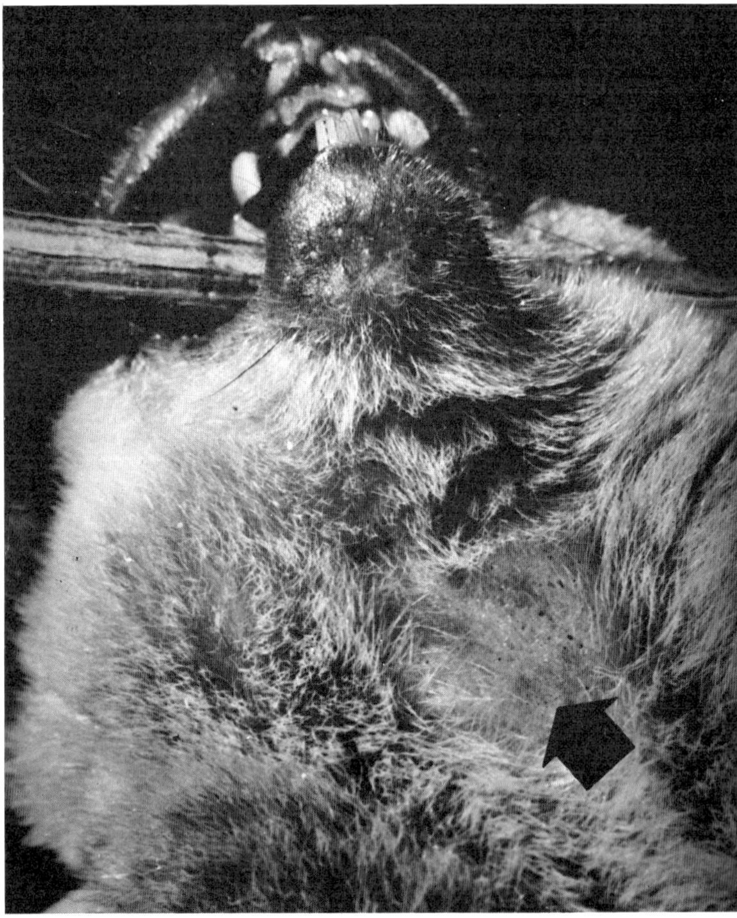

Fig. 4. External aspect of the cutaneous gland (arrow) under the neck of a male *Phaner furcifer* (photo by J.J. Petter).

miandra, 1971). In the male, this gland located below the chin, forms a humid and pink circular area of about 20 mm in diameter (Fig. 4). The female has only a whitish longitudinal bare patch of about 8 × 2 mm, without any evidence of secretory activity. The throat gland is involved in allomarking during bouts of allogrooming. Raising his head, the male rubs his throat in a typical movement on the head, shoulders, and back of the female (field and captivity observations). The marking of branches is exceptional, and was observed on only one occasion in the field (Petter *et al.*, 1971), and rarely in captivity.

Whereas vocalizations were always associated with the numerous bouts of territorial confrontations observed in the field, no marking of branches was observed as occurs in *Lemur* sp. in such situations (E. Pages, personal communication).

Visual Signals

Visual signals of *Phaner furcifer* appear to be quite similar to those of other nocturnal prosimians, particularly other Cheirogaleidae and Lorisidae (Charles-Dominique, 1974a). For example, the position on the branch (above, crouched, underneath), and eventually flight towards the ground are behaviors associated with dominance relations. The dark fork of the face is probably a signal of specific recognition.

Tactile Signals

Allogrooming is frequent between males, females, and juveniles. The protagonists engage in allogrooming face to face on the branch. They often hang upside down by their hindfeet from the branch, a typical position of Lorisidae and Cheirogaleidae (Charles-Dominique, 1977a; Pages, Chapter 4).

Male-Female Relations

Of the five adult males followed in their territories, one was a "bachelor," three were monogamous, and one bigamous. About 8 months after the mating period (June), these males continued to share many social relations with females. Equivalent situations are found in other nocturnal prosimians such as *Microcebus murinus* (Martin, 1972a,b; Perret, 1977), *Microcebus coquereli* (Chapter 4), *Lepilemur leucopus* (Charles-Dominique and Hladik, 1971), *Perodicticus potto, Galago demidovii, Galago alleni* (Charles-Dominique, 1972, 1977a,b), *Galago senegalensis* (Bearder and Martin, 1980). Whereas in Cheirogaleinae and Lorisidae, male–female relations occur during a short period of the night, more or less based on "visits" and olfactory messages (urine), relations are almost permanent in *Phaner furcifer*. The male and female are practically in continuous vocal contact throughout the night and are in close proximity for at least half of the night. Often, during movements, the male follows the female at a distance of 1–10 m. When the female arrives in a gum-producing Taly, the male waits 2–5 m behind, approaching to eat only when the female has left the tree. If the female continues farther on, the male may remain at a distance, but more often, 5–15 minutes later they renew vocal exchanges and the male rejoins the female. After being separated, the female was observed arriving in a Taly still occupied by the male, who immediately withdrew. The female seems to be "dominant" in this and most other situations.

Male–female and male–juvenile allogrooming has been observed on twelve occasions during the night. The marking of the female by the male throat gland may reinforce social ties. Generally the male and the female of a pair sleep in the same tree hole.

Monogamous Pairs (♂ 1 – ♀ 1, ♂ 2 – ♀ 2, ♂ 4 – ♀ 4)

The male usually follows the female or stays behind at a distance of 10–30 m, regularly exchanging vocalizations. On two occasions we observed ♀ 1 followed by ♂ 1, passing through the bachelor male's (♂ 5) territory. At the boundary, ♂ 1 halted progression and ♂ 5 approached ♀ 1. They engaged in allogrooming for about 1 minute, then resumed locomotion, ♂ 5 following ♀ 1 at 1–5 m of distance, while ♂ 1 remained at the boundary emitting numerous contact calls ("Hon . . . Hon . . . Hon"). On both occasions, ♀ 1 returned to ♂ 1's territory (after 30 or 45 minutes) and the situation reversed: ♂ 5 stopping at the border relayed by ♂ 1. These situations are rare and the case of ♀ 1–♂ 1–♂ 5 cannot be interpreted as true polyandrism. Generally male–female contacts (allogrooming, vocal exchanges, sleeping sites) are exchanged between the same individuals (♀ 1 and ♂ 1, ♀ 2 and ♂ 2, ♀ 4 and ♂ 4). In ♀ 2–♂ 2 and ♀ 4– ♂ 4 pairs, the male and female territory are perfectly superposed (see below).

Polygamous Group (♂ 6 – ♀ 6 – ♀ 7)

The territory of ♂ 6 overlaps most of ♀ 6 and ♀ 7 territories. The male–female relations are equivalent to those described above but ♂ 6 sometimes slept with ♀ 6 and her infant, and at other times with ♀ 7 and her infant. During the night he had simultaneous vocal exchanges with ♀ 6 and ♀ 7. He followed one of these two females, changing female from time to time during the night.

"Bachelor" Male (♂ 5)

This fully adult male lived alone, but occasionally had contacts with adjacent females which were, however, paired with adjacent males (♀ 1, ♀ 6, ♀ 7). He participated in vocal exchanges in the southeast "meeting area" and had been observed on two occasions courting the juvenile female ♀ 6′ (5-month-old infant of ♀ 6): ♂ 5 followed ♀ 6′ seeking contact, stopping his progression and crouching on the branches when ♀ 6′ turned back to groom him. On other occasion ♂ 5 was observed with ♀ 6, ♀ 7 and their infants in intense bouts of allogrooming.

After the experimental suppression of ♂ 4 and the extension of ♂ 5 territory into the previous ♂ 4 territory (see below), ♂ 5 engaged in courtship behavior with ♀4: pursuit of the female, seeking of contacts, anogenital flairing and crouching on the branches when the female turns back (typical nocturnal prosimian courtship behavior; Charles-Dominique, 1977a).

Territories and Territoriality

Mapping of all recorded itineraries yielded a network of lines of varying density: high density at the proximity of gum-producing trees and in overlapping areas ("meeting zones"), and low density in the other areas. Male and female territories are drawn separately since representation of the itineraries of all individuals would be unreadable. Each territory is shown by a line which encompasses the recorded itineraries (Figs. 5 and 6).

The mean surface area of territories is about 4 ha (4 ha for females and

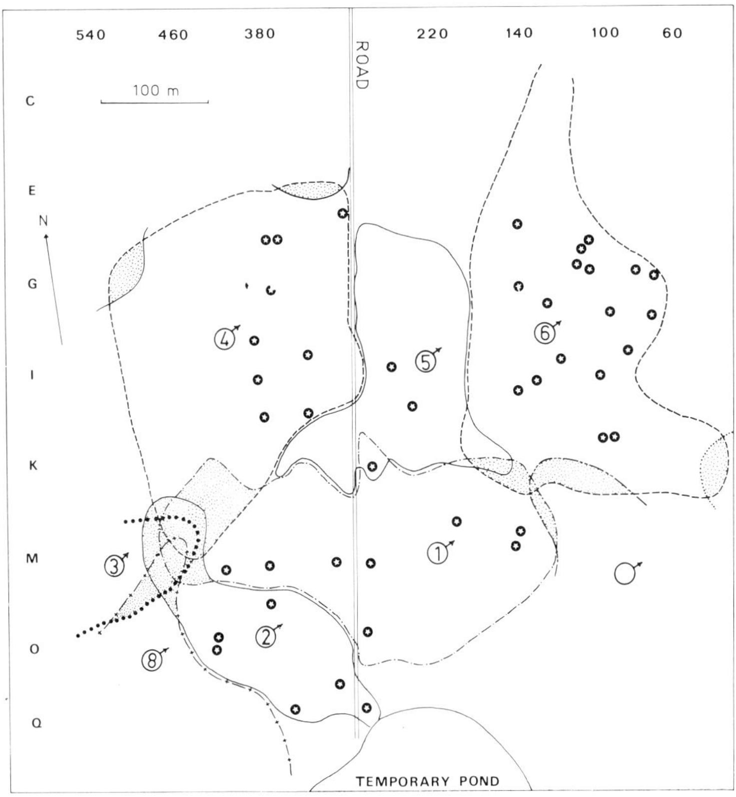

Fig. 5. Distribution of territories of male *Phaner furcifer* in the study area. Letters and numbers show localization of trails (see Chapter 1, Fig. 5 for detailed map of the study area). Gum-producing *Terminalia* trees are indicated by stars. Meeting areas are shaded.

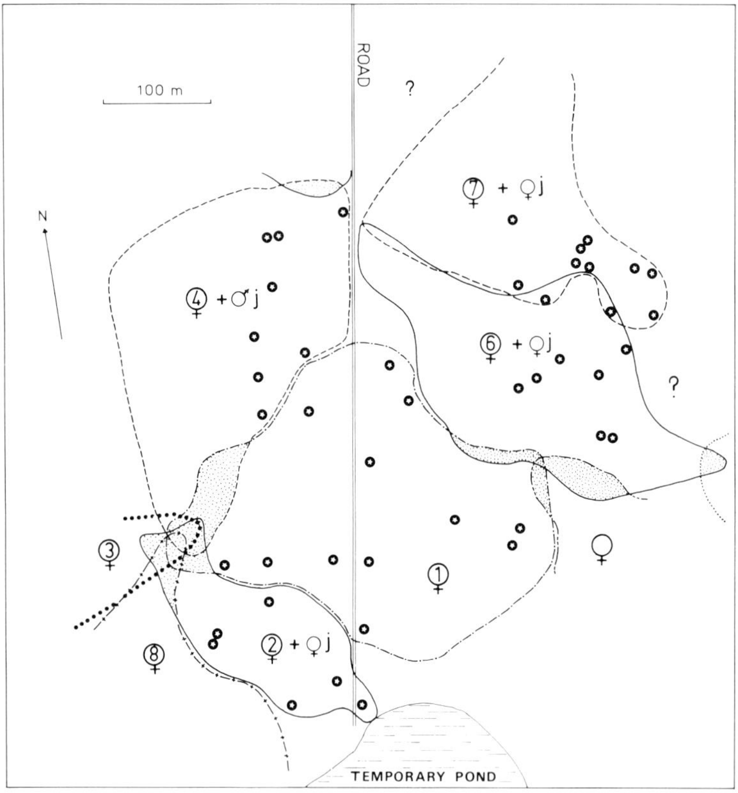

Fig. 6. Distribution of territories of female *Phaner furcifer* in the study area (localization as in Fig. 5). Juveniles occupy the same territory as the mother. *Terminalia* trees are indicated by stars. Meeting areas are shaded.

3.8 ha for males). In fact, the size of the territories appears to depend to a large extent on the distribution of Taly trees.

Male territories are generally adjacent to the territories of other males except in a limited overlapping zone (Fig. 5). A similar situation exists for females. The individual territory of a male–female pair are superposed. In the case of the paired females (♀1 ♀6 ♀7) close to the bachelor male, territories overlap that of ♂5 (Fig. 7), but the boundaries of consort males territories are adjacent to that of ♂5.

Overlapping Zones or "Meeting Areas"

In all Cheirogaleinae and Lorisidae species studied, territorial exchanges are exhibited in overlapping zones regularly visited and marked by the dif-

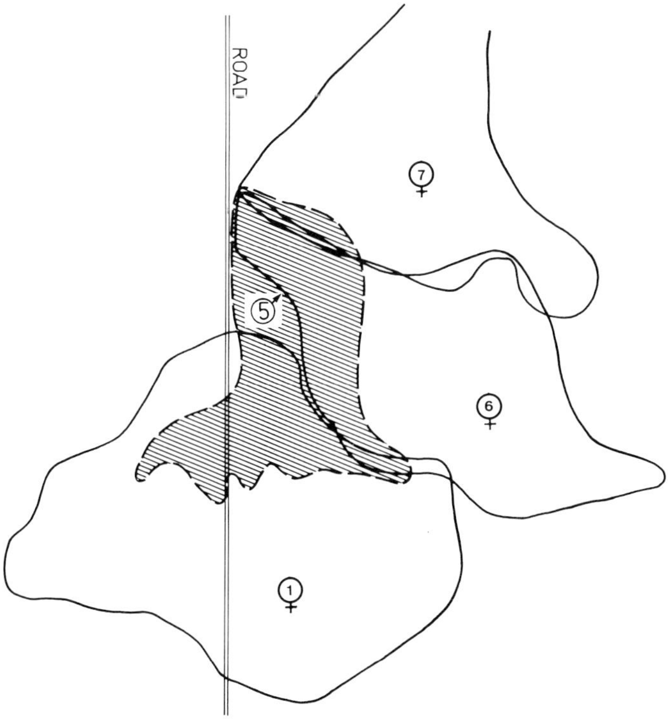

Fig. 7. Pattern of overlapping of three females' territories and that of one adult bachelor male (see Figs. 4 and 5). The limits of the females' territories are juxtaposed, but overlap with that of the male (shaded).

ferent protagonists whose visits rarely coincide temporally. By means of such reciprocal marking (usually urine) each individual can recognize the immediate neighbors, confirm their presence, and also inform them of its presence (= differed territorial communication, for example *Perodicticus potto, Galago alleni,* Charles-Dominique, 1974b, 1976, 1977a,b; *Microcebus coquereli,* Chapter 4). In *P. furcifer* the territorial organization is equivalent (overlapping zones) but scent marking is absent and vocalizations are predominant in social and territorial relations. During the night the different individuals of *P. furcifer* regularly emit series of "Kiu" calls which elicit other "Kiu" calls in the vicinity. An animal approaching an overlapping zone often emits numerous calls. Generally several neighbors join this animal in the overlapping zone ("meeting area") and, at a few meters distance, all emit simultaneously numerous "Kiu" calls, as well as male calls, in a concert which may last for about 10–20 minutes. These concerts generally involve 3–5 individuals, sometimes as many as 9 (3 individuals, 3 observations; 4 individuals, 5 observations; 5 individuals, 2 observations; 6 individuals, 2 observations; 7 individuals, 1 observation; 9 individuals, 1 observation). Most of these concerts occur after 20.30 hours. In Fig. 8 the increase in the frequency of calls between 21.00 and 00.00 hours corresponds to the maximum of concert calling in the meeting areas. In spite of the high intensity of the calls, the animals remain calm and can approach one another at a distance as close as 10 cm without fighting.

Meeting areas occupy 1/25 of the study area; no Taly tree were noticed in these zones.

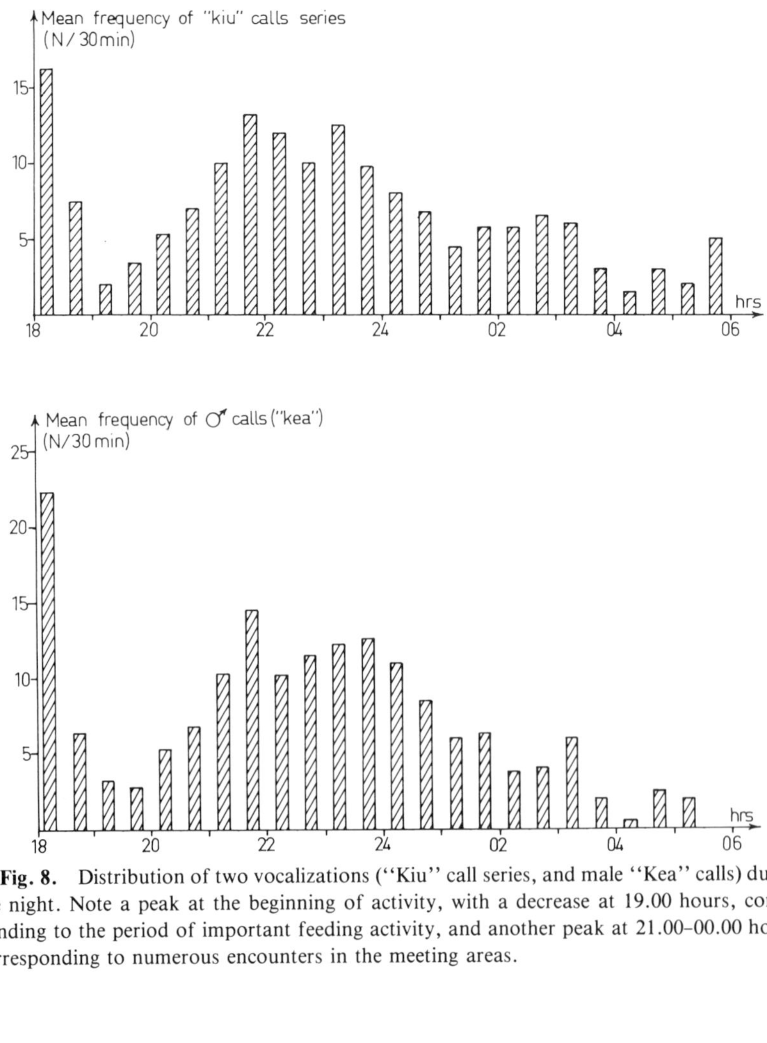

Fig. 8. Distribution of two vocalizations ("Kiu" call series, and male "Kea" calls) during the night. Note a peak at the beginning of activity, with a decrease at 19.00 hours, corresponding to the period of important feeding activity, and another peak at 21.00–00.00 hours, corresponding to numerous encounters in the meeting areas.

Experimental Suppression of a Paired Male (♂4)

The following experiment in June 1974, which consisted of the experimental suppression of one male, illustrates some of the mechanisms involved in maintaining social relations in *Phaner furcifer:*

On June 9, ♂4, paired with ♀4, was captured at 18.34 hours and caged in the center of his territory (Fig. 9). At 20.15 hours, his neighbor, bachelor ♂5 entered his territory emitting numerous male calls, and "Kiu" calls. The itinerary of ♂5 was marked by a series of swings to and from the previous boundary; ♂5 emitted numerous calls to which the neighbors answered, particularly ♂4 in the cage. Observations ended at 01.26 hours.

On June 10, ♂5 who slept in his previous territory, entered ♂4's territory at twilight. At 19.00 hours ♂4 was placed at a distance of 4 km so that his call would not influence ♂5's behavior. ♂5 continued his incursions and vocalizations as on the previous day. Observations ended at 00.00 hours.

On June 11, ♂5 slept in ♂4's previous territory (Fig. 10). Immediately after awakening, ♂5 went towards ♂1's boundary emitting numerous male calls. ♂1 entered ♂4's previous territory (18:08 hours; ♂5 and ♂1 fought and fell to the ground, then ♂5 pursued ♂1 who returned into his

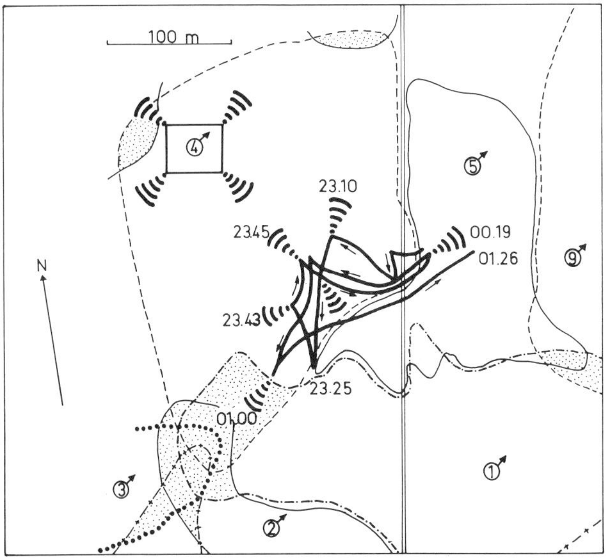

Fig. 9. Movements of ♂5 (bachelor) after capture of neighboring ♂4 in his territory. ♂4 caged in his territory exchanged numerous vocalizations with ♂5 who made incursions (heavy line) into the border of ♂4's territory. Calls and time of emittance are indicated by the "wave" symbol.

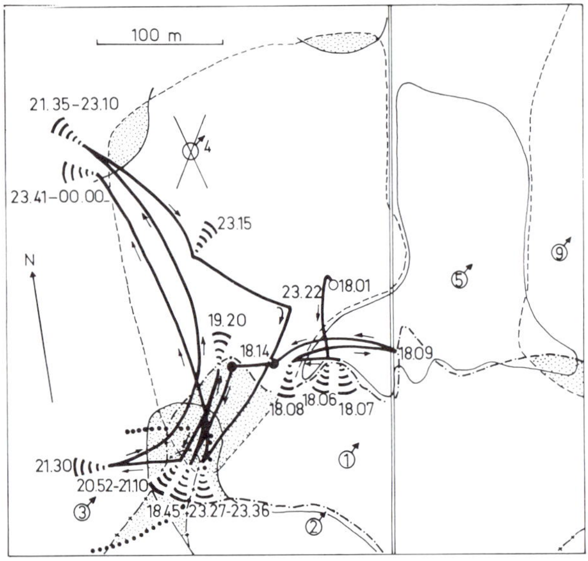

Fig. 10. Movements of ♂5 after ♂4 was removed to a distance of 4 km. In this situation ♂5 penetrates within the entire territory of ♂4, and rejects at the border the neighboring males which also try to penetrate the territory. Symbols as in Fig. 9.

own territory (18:09 hours). ♀1 came toward ♂5 (18:14), they groomed each other for 6 minutes, then ♀1 returned toward ♂1. ♂5 went to the south meeting area where numerous *Phaner* were present and emitted many more vocalizations than usual. Until 00.00 hours (end of observations) ♂5 passed successively from the north meeting area to the south meeting area emitting numerous male calls and "Kiu" calls. The two Taly trees where ♂5 habitually fed were fitted with electrical switches connected to a graphic recorder: only one visit occurred at 05.10 hours, just before sunrise, in comparison to 4–5 visits during the previous week.

On June 12, ♂5 occupied his previous territory plus ♂4's territory. The calmness had returned to the study area. ♂5 was observed courting ♀4 (the previous ♂4's female).

This experiment shows how vocalizations can inform the neighbors of the presence of an occupant of a given territory. However, vocalizations alone are not sufficient to maintain territorial stability since small incursions of ♂5 occurred when ♂4 was caged in the center of his territory and emitted calls. It appears that vocalizations throughout the territory, plus encounters and vocalizations in the meeting areas, are both necessary for the maintenance of territorial stability. A similar type of experiment conducted in a natural population of *Lepilemur leucopus* (Charles-Dominique and Hladik, 1971) yielded very comparable results. In both cases the actual presence and vocalizations of the normally established animal were necessary to maintain the integrity of the territorial boundaries, even on a short-term basis.

DISCUSSION

If many Strepsirhines have been observed occasionally to ingest some quantity of gums, *Phaner furcifer* is particularly specialized for this type of diet. This situation is similar to that of the African needle clawed bushbaby, *Galago (Euoticus) elegantulus,* and the two species share many other features: a long and narrow tooth comb, long extensible tongue, well-developed upper first premolar, large cecum, sharpened nails facilitating climbing on large branches and trunks, fixed itineraries associated with exploitation of gum trees, similar size, dark dorsal stripe and certain similar vocalizations. Except for the last two characters, all are related to the exploitation of gum resources. However, there are significant differences in the social organization of these two species. Whereas the social organization of *G. elegantulus* is of the Galagidae and Cheirogaleinae type (social communications based upon urine marking and vocalizations, solitary activity during the night, and matriarchies, Charles-Dominique, 1977a), *P. furcifer* represents a particular case of "pregregarious" social organization (absence of deferred urine marking, predominance of vocal exchanges, and relative synchronization of pair movements during the night). These divergences illustrate that social organization is not exclusively the result of ecological pressures (Crook, 1970) but, as suggested by Eisenberg *et al.* (1972) and Charles-Dominique (1978), may also be the consequence of a combination of phylogenetic characters and ecological adaptations. Thus, different social solutions can be adopted for similar ecological niches.

Among primates all diurnal species are gregarious and almost all nocturnal species are solitary [23 solitary species, and only three gregarious species: *Aotus trivurgatus* (Moynihan, 1976), *Lemur mongoz* (Tattersal and Sussman, 1975; Petter *et al.,* 1977), *Hapalemur griseus* (Petter *et al.,* 1977)]. The pregregarious organization as seen in *Phaner furcifer* represents an intermediate stage between these two modes of social organization and the emphasis on vocal communications is probably one of the conditions associated with the transition to gregariousness.

REFERENCES

Bearder, S.K., and Doyle, G.A. (1974). Ecology of Bushbabies, *Galago senegalensis* and *Galago crassicaudatus,* with some notes on their behavior in the field. *In* "Prosimian Biology" (R.D. Martin, G.A. Doyle, and A.C. Walker, eds.), pp. 109–130. Duckworth, London.

Bearder, S.K., and Martin, R.D. (1980). In preparation.

Charles-Dominique, P. (1971). Eco-éthologie des Prosimiens du Gabon. *Biol. Gabonica* **7**, 121–128.

Charles-Dominique, P. (1972). Ecologie et vie sociale de *Galago demidovii* (Fisher 1808, Prosimii). *Z. Tierspsychol., Suppl.* **9**, 7–41.

Charles-Dominique, P. (1974a). Aggression and territoriality in nocturnal prosimians. *In* "Primate Aggression, Territoriality and Xenophobia" (R. Holloway, ed.), pp. 31–48. Academic Press, New York.

Charles-Dominique, P. (1974b). Vie sociale de *Perodicticus potto* (Primates, Lorisidés). Etude de terrain en forêt équatoriale de l'Ouest africain du Gabon. *Mammalia* **38**, 355–379.

Charles-Dominique, P. (1976) Leg gommes dans le régime alimentaire de *Coua cristata* à Madagascar. *Oiseau R.F.O.* **46**, 174–178.

Charles-Dominique, P. (1977a). "Ecology and Behaviour of Nocturnal Primates." Duckworth, London.

Charles-Dominique, P. (1977b). Urine marking and territoriality in *Galago alleni.* A field study by radio-telemetry. *Z. Tierspsychol.* **43**, 113–138.

Charles-Dominique, P. (1978). Solitary and gregarious prosimians: Evolution of social structures in Primates. *In* "Recent Advances in Primatology" (D.J. Chivers and K.A. Joysey, eds.), Vol. 3, pp. 139–149. Academic Press, New York.

Charles-Dominique, P., and Hladik, C.M. (1971). Le Lépilémur du sud de Madagascar: écologie, alimentation et vie sociale. *Terre Vie* **25**, 3–66.

Coimbra-Filho, A. (1972). Aspectos Inéditos do comportamento de Sagüis do gênero *Callithrix* (Callithricidae, Primates). *Rev. Bras. Biol.* **32**, 505–512.

Crook, J.H. (1970). The socio-ecology of Primates. *In* "Social Behavior in Birds and Mammals" (J.H. Crook, ed.), pp. 103–166. Academic Press, New York.

Eisenberg, J. F., Muckenhirn, N. A., and Rudran, R. (1972). The relation between ecology and social structure in Primates. *Science* **176**, 863–874.

Hausfater, G., and Bearce, W.H. (1977). Acacia tree exudates: Their composition and use as food source by Baboon. *East Afr. Wildl. J.* **15**.

Hill, W.C.O. (1953). "Primates, Comparative Anatomy and Taxonomy," Vol. 1. Edinburgh Univ. Press, Edinburgh.

Martin, R.D. (1972a). A preliminary field study of the Lesser Mouse Lemur *(Microcebus murinus* J.F. Miller 1777). *Z. Tierpsychol., Beih.* **9**, 43–89.

Martin, R.D. (1972b). Review Lecture adaptative radiation and behaviour of the Malagasy lemurs. *Philos. Trans. R. Soc. London, Ser. B* **264**, 295–352.

Moynihan, M. (1976). "The New World Primates: Adaptive Radiation and the Evolution of Social Behavior, Languages, and Intelligence." Princeton Univ. Press, Princeton, New Jersey.

Pariente, G.F. (1974). Influence of light on the activity rhythms of two Malagasy lemurs: *Phaner furcifer* and *Lepilemur mustelinus leucopus. In* "Prosimian Biology" (R.D. Martin, G.A. Doyle, and A.C. Walker, eds.), pp. 183–198. Duckworth, London.

Perret, M. (1977). Influence du groupement social sur l'activité sexuelle saisonnière chez le mâle *Microcebus murinus* (Miller 1777). *Z. Tierpsychol.* **43**, 159–179.

Petter, J.J., and Charles-Dominique, P. (1978). Vocal communication in Prosimians. *In* "The Study of Prosimian Behavior" (G.A. Doyle and R.D. Martin, eds.), pp. 247–305. Academic Press, New York.

Petter, J.J., Schilling, A., and Pariente, G. (1971). Observations éthologiques sur deux lémuriens malgaches nocturnes, *Phaner furcifer* et *Microcebus coquereli*. *Terre Vie* **3**, 287–327.

Petter, J.J., Albignac, R., and Rumpler, Y. (1977). Mammifères Lémuriens (Primates Prosimiens). *Faune Madagascar* **44.**

Petter-Rousseaux, A. (1962). Recherches sur la biologie de la reproduction des Primates inférieurs. *Mammalia* **26,** Suppl. 1, 1–88.

Petter-Rousseaux, A. (1968). Cycles génitaux saisonniers des lémuriens malgaches. Cycles génitaux saisonniers de Mammifères sauvages. *Entretiens Chize, Ser. Physiol.* **1,** 11–18.

Petter-Rousseaux, A. (1974). Photoperiod, sexual activity and body weight variations of *Microcebus murinus* (Miller, 1777). *In* "Prosimian Biology" (R.D. Martin, G.A. Doyle, and A.C. Walker, eds.), pp. 365–373. Duckworth, London.

Rumpler, Y., and Andriamiandra, A. (1971). Etude histologique des glandes de marquage de la face antérieure du cou des Lémuriens malgaches. *C.R. Seances Soc. Biol. Ses Fil.* **165,** 436.

Sauer, E.G.F., and Sauer, E.M. (1963). The South West African bushbaby of the *Galago senegalensis* Group. *J. South West Afr. Sci. Soc.* **16,** 5–36.

Schilling, A. (1978). The possible role of urine in territoriality of some nocturnal prosimians. *Symp. Zool. Soc. London* (in press).

Tattersall, I., and Sussman, R.W. (1975). Observation on the ecology and behavior of the mongouse lemur (*Lemur mongoz mongoz,* Linnaeus, Primates, Lemuriformes) at Ampijoroa, Madagascar. *Anthrop. Pap. Am. Mus. Nat. Hist., New York* **52,** No. 4, 193–216.

4 *Ethoecology of Microcebus coquereli during the Dry Season*

E. PAGES

INTRODUCTION

The pronounced seasonal cycle of the forest on the West Coast of Madagascar, to which the nocturnal prosimian species are subjected, has resulted in behavioral and physiological adaptations which are, in certain cases, peculiar for primates (see Chapters 2 and 3). The adaptations of the different species are most evident during the dry season of austral winter, when food resources are greatly reduced. In the case of *Microcebus coquereli,* these adaptations are mainly behavioral particularities related to social organization and nutritional habits. Although active throughout the year, *M. coquereli* enters into a period of sexual inactivity (as do other Malagasy vertebrates; Petter-Rousseaux, 1968, 1974; Chapter 6) during austral winter.

Microcebus coquereli has a geographical distribution restricted to the West Coast of Madagascar, except for the extreme south (Petter *et al.,* 1977). This species has only been the object of one previous field study (Petter *et al.,* 1971) and was poorly known even by the local population.

The present field study was conducted mainly in June and July 1974 during which observations were made on diet, habitat utilization, activity rhythms, and social behavior. Additional data for comparative purposes were provided by the other field workers of the "Morondava Programme" (P. Charles-Dominique; C. M. Hladik; G. Pariente; J. J. Petter).

The successful reproduction of *M. coquereli* in the animal house of Brunoy permitted supplementary detailed observations.

STUDY AREA AND METHODS

The study area, in the Marosalaza forest (50 km north of Morondava), has been described by A. Hladik (Chapter 1). In this forest, precipitation is very frequent during the warm months of December/January/February (austral summer) and, accordingly, food resources are abundant (Chapter 2). In contrast, the period of June/July/August (austral winter) is dry and relatively cold, with food resources becoming progressively scarce. *Microcebus coquereli* were mainly observed in the southern part of this area (Trails M to W, see Fig. 5, Chapter 1) near the temporary pond, where the forest is more dense and animal density high.

Microcebus coquereli were trapped according to the methods of Charles-Dominique (Charles-Dominique and Bearder, 1978), using spring-loaded cages. In addition, during the daytime, some individuals were captured in the nest with a sliding noose attached to a long pole.

Twenty-one individuals were consistently identified in the study area, and six were equipped with radio transmitters allowing rapid location (Zimmermann *et al.,* 1976). Transmitter and batteries were embedded in paraffin wax protected by a thin seal of acrylic resin, and attached around the neck of the animal by a plastic collar in which the antenna was included. The animals rapidly became accustomed to this transmitter. Individual locations were recorded every 15 minutes and in some cases continually (especially during rapid movements). In general, one animal was followed at a time, or, in the case of a mother and her young, or of a male and female pair in the same zone, both individuals were tracked alternatively on short successive periods. Observations lasted at least half the night. However the entire night was covered at least once a week, and when a newly captured animal was released.

FEEDING BEHAVIOR

During the austral summer, *M. coquereli* feed on a large variety of food resources: insects, spiders, frogs, chameleons, fruits, flowers, buds, gums, and insect secretions. Secretions of homopteran larva play a major role (Petter *et al.,* 1971; Chapter 2) and, in June (beginning of the austral winter) accounted for up to 50% of the feeding observations. At the larval stage, the colonial insects exude a sweet secretion when disturbed. The droplets of exudate fall on the leaves and dry to form a hard white concretion. *Microcebus coquereli* regularly visit several colonies each night and lick the liquid secretion directly from the insects, as well as the white crystallized "sugar" which is collected by the hand and the tooth comb.

Prey behavior is well developed, and the adults have often been observed to capture large insects, small birds, and mammals, and also to eat eggs. During austral winter, *M. coquereli* often begins feeding activity by first

TABLE I

Comparison of the Densities of *Microcebus coquereli*, Homopteran Colonies, and Free Living Insects [a]

	Area 1 (Trail A to M)	Area 2 (Trail M to R)	Area 3 (Trail R to W)
Number of *M. coquereli* captured in 15,000 m^2	2	4	9
Number of recent nests of *M. coquereli* counted along 1 km of transect	2	5	13
Number of colonies of homopteran larvae counted along 1 km of transect	42	11	7
Number of *Buxus madagascariensis* with leaves counted in 100 m^2 along transects	6	8	12
Dry weight of the free living insects gathered from 5 kg of fresh foliage of *Buxus madagascariensis* (gm)	0.5	0.5	1

[a] The density of *Microcebus coquereli* seems to be correlated with that of free living insects, and not with the actual number of homopteran larvae present in different areas of Marosalaza forest.

visiting homopteran colonies, after which insects may be hunted in the dense foliage of some evergreen trees (Fig. 1A) such as *Buxus madagascariensis* (29),* *Strychnos decussata* (93), *Grewia* sp. (166/174).

The animal also occasionally forages at ground level. Gums and other exudates may also be collected on *Terminalia* sp. (48/49) and *Adansonia* sp. (20/22).

In spite of the fact that feeding behavior is mainly devoted to collection of homopteran secretions, the distribution of *M. coquereli* appears to depend on the availability of free living insects, rather than that of homopteran larvae colonies, as suggested by a survey of the study area (Table I). This is explained by the extremely low protein content (1.4%) of the insect secretion which necessitates the protein complement provided by insects (Chapter 2).

* The numbers in parentheses refer to the list of plant species (Chapter 1, Appendix I).

Fig. 1. *Microcebus coquereli* feeding and nest building: (A) One of the animals equipped with a radio transmitter foraging at night in a zone of dense foliage of the Marosalaza forest. (B) A nest built in the lianas in the Marosalaza forest. The animal (circle) chased during daytime, stays just above the nest. (C) A nest built in captivity showing interlacing of lianas and branches, and the animal at the entrance. (D) The animal carrying a small branch during nest building.

NEST BUILDING BEHAVIOR

In nature as well as in captivity, *M. coquereli* constructs a large spherical nest which measures 50 cm in diameter (Fig. 1B and C). This nest is made of lianas, branches, and leaves which are interlaced, and usually located in the fork of a large branch or among dense lianas. Small branches are broken by use of the teeth and simultaneously held with the hands. The branch held in the mouth is transported to the nest (Fig. 1D), and wound around the supporting branches toward the interior using the hands. The free end is then fixed between the other interlaced branches. Final trimming is done with the teeth by cutting lianas emerging from the sphere and interlacing the free ends. In captivity, male, female, and young, all participate in nest construction.

In nature, each individual may utilize up to 12 nests which are successively occupied, but about half of them seem to be abandoned nests in bad repair which may occasionally be used. Some of these nests are also used by *Phaner furcifer* (Petter *et al.*, 1971). The nests are habitually located at a height of 2–10 m, often in trees which do not shed their leaves during the dry season (e.g., Euphorbiaceae) and are thickly covered with lianas, or occasionally may be built much higher in large bare trees.

NOCTURNAL ACTIVITY AND CIRCADIAN RHYTHMS

During the study period, there was normally little variation in the time of exit and return to the nest. The first sign of activity could be detected by variations in the radio signal, due to body movements while the animal was still in the nest. This generally occurred between 17:15 and 18:15 hours. The actual exit from the nest was between 17:45 and 18:45 hours. Upon leaving the nest, the animal usually remains in the near proximity, self grooms, and begins to move only when the light level is very low (Pariente, 1974; Chapter 5).

The return to the nest occurred between 5:40 and 6:05 hours, just before dawn. The lunar cycle does not appear to affect activity rhythm. However, low temperatures may have an important effect: the exit from the nest occurred later on cold nights (latest observed: 22:00 hours), and return to the nest was earlier (earliest observed: 22:30 hours).

Rest periods lasting for one hour and more are frequent, especially during the first half of the night. These periods of inactivity may be longer during cold nights, and in this case the reduction in activity is usually at the expense of social activities rather than of feeding and other solitary activities (see following discussion and Fig. 9).

The different types of activity take place within different areas of the home range (Fig. 5A and B), and at different heights, as illustrated by Fig. 2. This vertical distribution of the various activities is roughly correlated with the architecture of the vegetation including an upper level of large and usually bare trees (at the end of winter), a middle level which is rather discontinuous, and a lower level composed of bushes and small trees in which the foliage remains relatively dense even in austral winter.

Different patterns of locomotion have been observed:

1. Slow movements in various directions during feeding activity, which permit close examination of dense foliage where insects may be found, usually at heights of 1.5 to 3 m.
2. Rapid locomotion occurs at medium height (2–5 m) and includes short leaps and running, mainly in the central area of the home range. In one hour, 500 m can be covered, and a total of 1000–1500 m during the entire night.
3. Locomotion along the border of the home range consists of rapid movements in a straight line, but with less leaps and usually at a lower elevation from the ground than in the preceding case.
4. Home range surveillance consists of displacement at great heights in the center of the home range, in very frequented areas located in the sparce foliage of large trees. These movements are frequently interrupted by stops.

Although the movements of an animal vary from one night to the other, a characteristic evolution of behavior during the night is evident (and is maintained in captivity as well). The night can be divided into two complementary phases. During the first half of the night (before about midnight or 1:00), the major activity is devoted to solitary behaviors (such as feeding), while the second half is dominated by encounters, vocalizations, and social activities.

SOCIAL ORGANIZATION

Home Range Utilization

Among the 21 individuals identified, six animals were regularly followed by radio-tracking (♂1, ♂6, ♀2, ♀10, juvenile ♂11, juvenile ♂12). The first animal (♂1), equipped with a radio transmitter in November 1973 by Charles-Dominique and C. M. Hladik, was relocated in the same area in June 1974. The animals were followed for 15 to 30 days, and a summary of all observations is presented in Fig. 3.

Fig. 2. Schematic cross section of the forest and localization of the main activities of *Microcebus coquereli* according to height and vegetation type: F, Feeding behavior; N, Nesting; P, Play and social contacts; R, Rest; S, Surveillance; V, Vocalization (single note call).

Six other animals were captured, marked, and recaptured, certain of them on several occasions (♂4, ♂5, ♂9, ♀3, ♀7, ♀8). At least eight other animals could be regularly observed and identified by their particular physical characteristics (♂13, ♂14, ♀15, ♀16, ♀17, juvenile 18, juvenile 19, juvenile 20).

The home range of adults includes a central area of relatively small size (15,000 m^2 for males; 25,000–30,000 m^2 for females) which is intensively utilized. This is surrounded by a large peripheral area (maximum observed 40,000 m^2 for males, and 45,000 m^2 for females) which is less frequently visited, with a less stable border, but probably requires several months of observation to be accurately determined. Overlapping of the central area may concern a female and her youngs (for example ♀10, juvenile ♂11, and juvenile ♂12, Fig. 3) or male and female (♀2 with ♂6, Fig. 4). There is a much greater degree of overlapping of the peripheral area, which permits social contacts between individuals of both sexes, even for distant neighbors.

A more detailed study of ♂6 provides a clear example of the significance of home range utilization (Fig. 5A and B). A total of 570 recordings of position and immediate observation of behavior permitted a precise picture of activity localization which allow the two main areas previously mentioned to be defined more accurately. The animal spent 80% of the night in the central area, even though the area represents only 30% of the total home range. Furthermore, 80% of the total activity in this area occurred between 18:00 and 0.00 hours.

As can be seen in Fig. 5A, the central area is characterized by solitary ac-

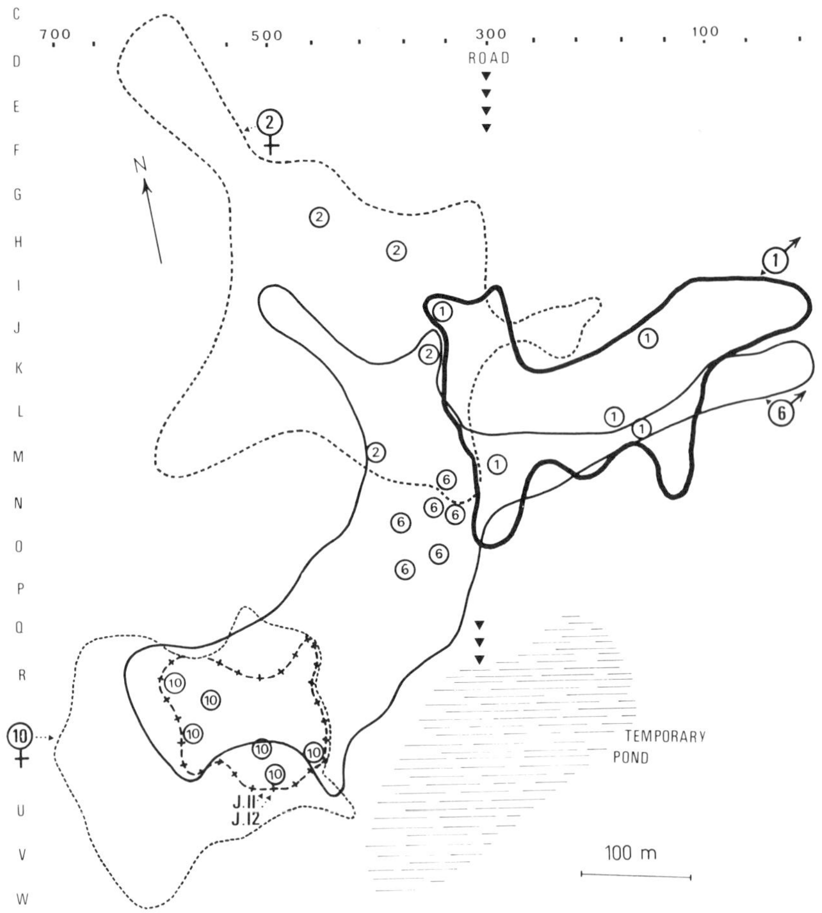

Fig. 3. Home ranges of six individual *Microcebus coquereli* followed by radio tracking. Solid lines show ranges of males; dashed lines show ranges of females; crosses show the ranges of two juveniles. The nests of different individuals are located in numbered circles. The trails of the study area are located by numbers and letters corresponding to Fig. 5, Chapter 1.

tivities such as feeding, self grooming, surveillance, rest, and possibly marking. The sudden release of a characteristic strong odor was detected when the animal was followed at close distance. It is likely that these emissions of specific odor participate in the marking of the central area and could play a role in territorial behavior (see below).

The northern part of the central area of ♂6 home range is shared with ♀2. Most of the encounters between these two animals occurred in this area between 22:00 and 2:00 hours.

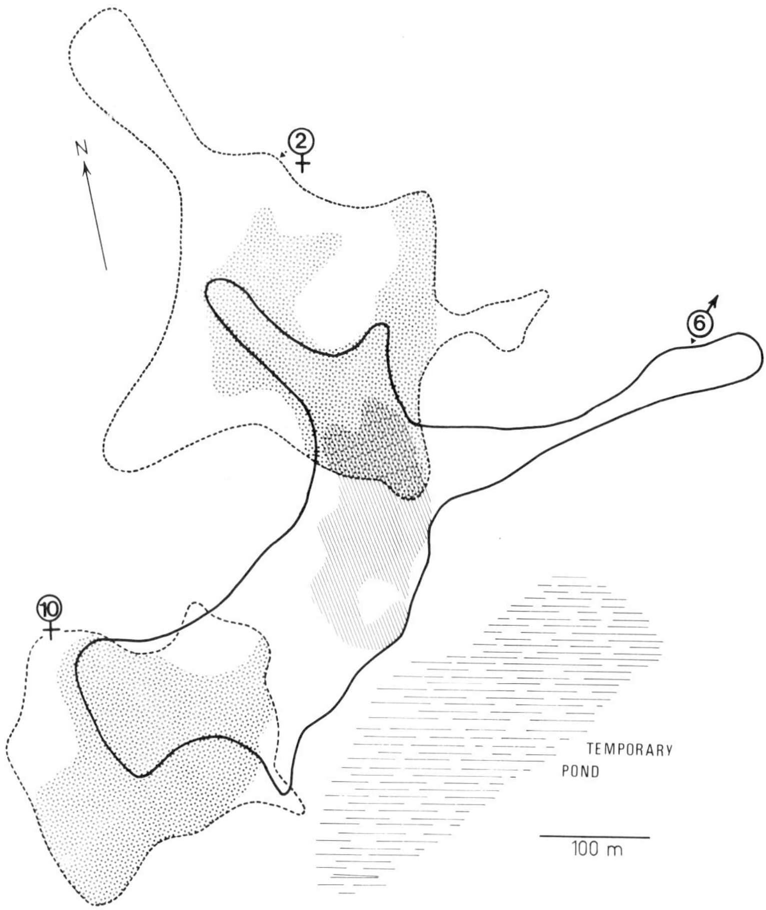

Fig. 4. Illustration of differential utilization of the home range by Male 6 and two neighboring adult females. Peripheral zones of home ranges show large overlaps between any two neighbors of opposite sex. In contrast, central zones (shaded) only show overlap between certain individuals linked by pair bonds or parental bonds.

The peripheral area, which represents about 70% of the total home range, accounted for only 20% of the observations of ♂6 (Fig. 5B). This area was frequented mainly after midnight. Feeding occurred on only one occasion, whereas social contacts with other individuals (♀2, ♀3, ♂1, ♂13, juvenile ♂20, juvenile ♂11, juvenile ♂12, and possibly ♂14, and ♀15) were often observed. The males generally make longer incursions into distant areas than females, and thus encounter more individuals in one night.

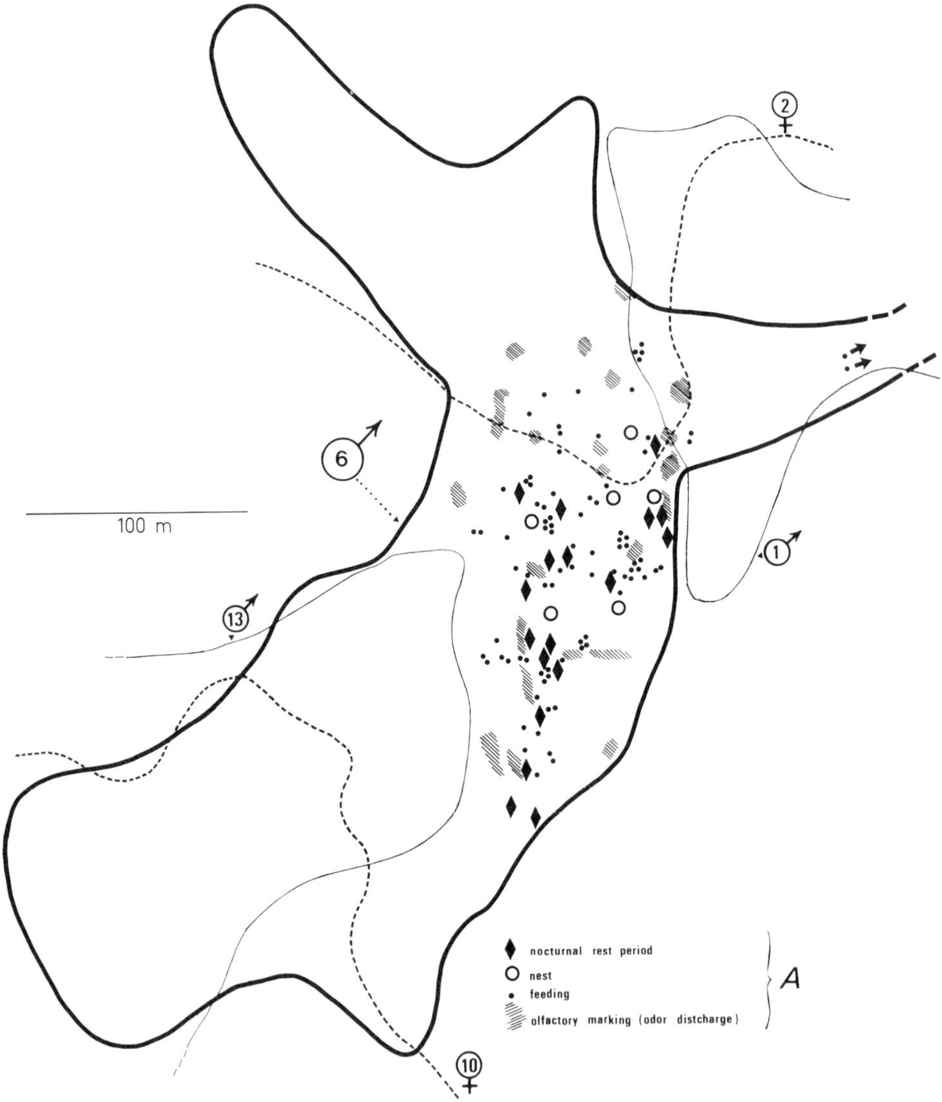

Fig. 5. Distribution of the activities of Male 6 in his home range, at night (100 hours of observation): (A) Solitary activities take place in the central zone of the home range and occur during the beginning of the night. (B) Social behaviors are mainly located in the overlapping zones of the home range and take place during the second half of the night.

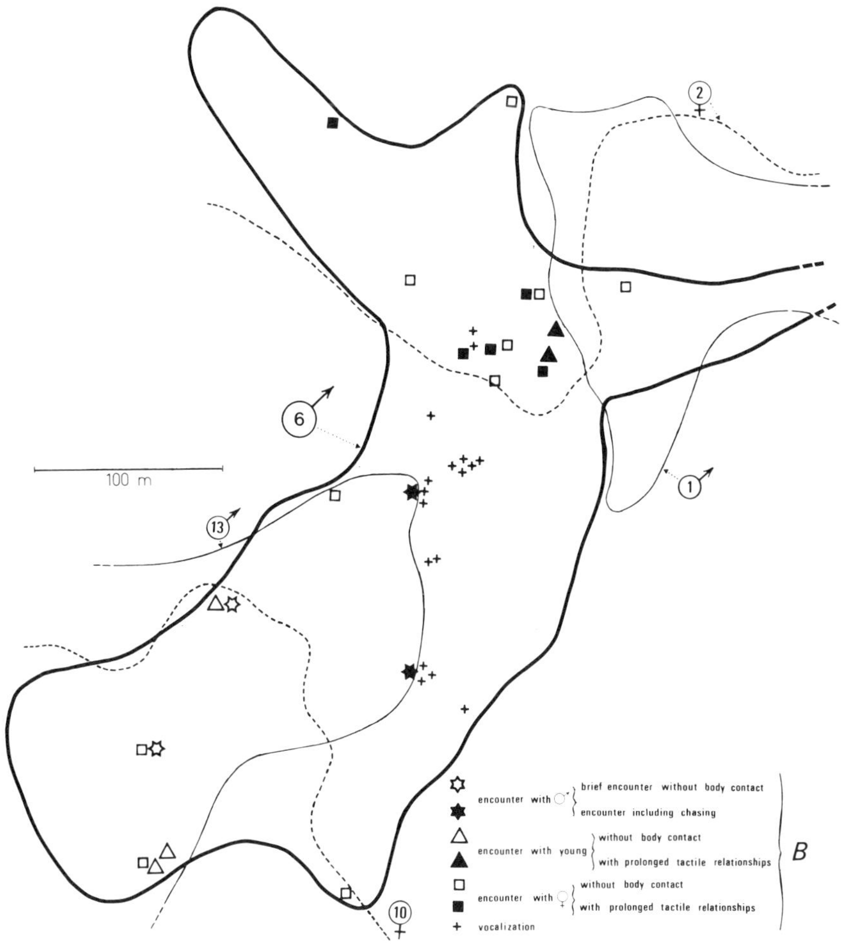

Fig. 5. (*Continued*)

The limits of the central area may be roughly determined within one night, when the animal is released after capture. On the contrary, even after more extensive observations (100 hours for ♂6; Fig. 6), the total extent of the peripheral area is probably only partially elucidated. This is due to the fact that, at the time of observation (season of sexual inactivity), deep penetrations into the neighboring home ranges were only occasional, whereas captures made during the austral summer indicated more movements. In fact, observations in captivity (Chapter 7) showed that more frequent social contacts occur during spring.

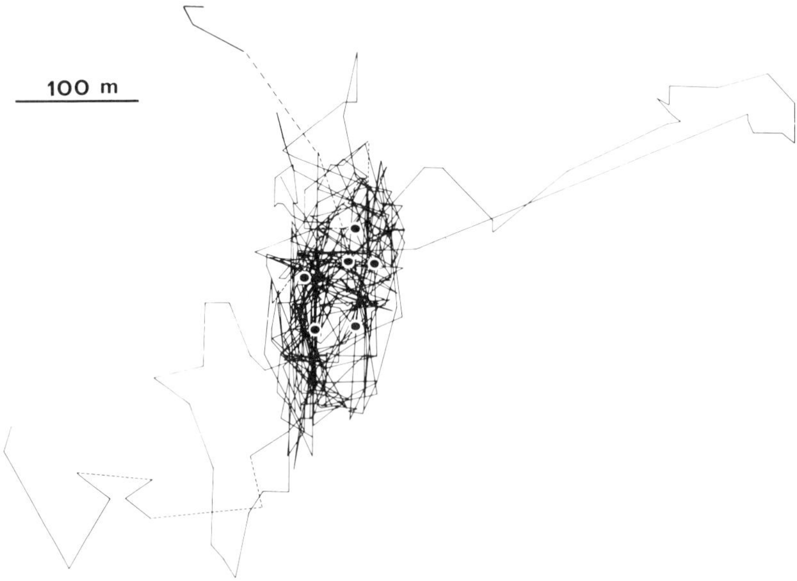

Fig. 6. Movements of Male 6 recorded during 100 hours of observation. The central area of the home range can be easily distinguished, while the peripheral zone is probably only partly assessed. Nests actually used are indicated by circles.

INTERCOMMUNICATION

Vocalization

Like other prosimians, *M. coquereli* uses a variety of vocalizations. For sonographic descriptions of these calls and comparison with those of other prosimians, see Petter and Charles-Dominique (1978). The most characteristic patterns are presented in Fig. 7. According to the acoustical characteristics and the behavioral context, eight types of calls can be defined:

1. The "contact call" resembles a "hum" or a "hein." It is the most frequent call and accompanies all movements and meetings between familiar individuals.
2. The "mother–infant meeting call" is a series of short modulated whistles. This call is generally emitted by mother and infant at dawn, before return to the nest. Loudness, tonality, and repetition rate vary with the context (Fig. 7A, B, C).
3. The "female single note call" resembles a "pfiou." This call elicits response in other females which may imply a role in territoriality.
4. The "male single note call" is a short intense whistle, emitted by

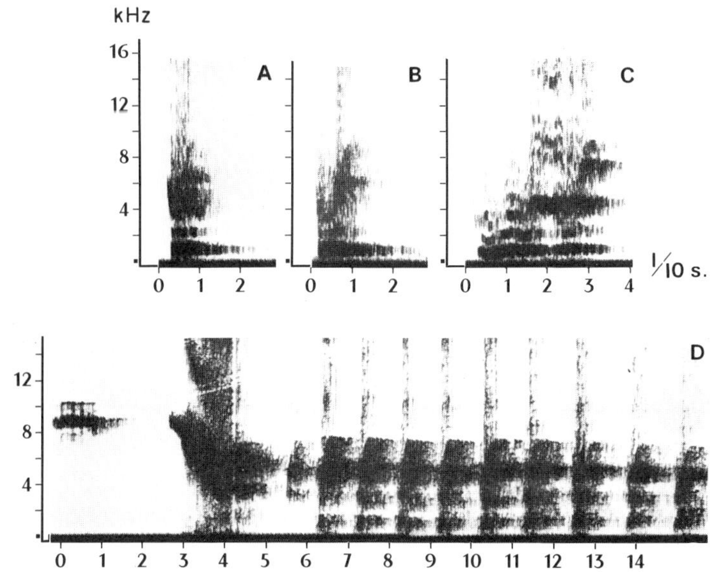

Fig. 7. Some characteristic vocalizations of *Microcebus coquereli* (Sonogram, filter 300 Hz wide). (A) Mother–infant meeting call (emitted by female). (B) Mother–infant meeting call (emitted by infant). (C) Same type of call, but of longer duration. (D) Sexual call sequence.

the male usually from a tall tree in the central part of the home range. Responded to by other males, as for the former call, it probably also has territorial significance.

5. The "alarm call" is a "croak" resembling the alarm call of other nocturnal prosimians.

6. The "agonistic call" is a short note rapidly repeated "tisk-tisk-tisk" emitted by young and adults of both sexes.

7. The "waking call sequence" is a series of notes progressively decreasing in intensity, emitted by females just after leaving the nest.

8. The "sexual call sequence" is a long characteristic sequence beginning with a weak whistle generally followed by several amplitude modulated notes, a hoarse "brroak" (Fig. 7D), emitted by both male and female adults during two or three days at the moment of estrus.

Marking Behavior

Marking behaviors of *M. coquereli* have been described by Schilling (1978) and compared to those of other prosimian species. The first type, punctuated urine marking (Fig. 8A), involves mainly urine, whereas in anogenital dragging (Fig. 8B), other substances (feces and glandular secre-

Fig. 8. Marking behavior of *Microcebus coquereli* in captivity. (A) Female performing punctuated urine marking by light contact of the clitoris. (B) Female marking by anogenital dragging (this type of marking exists in both sexes). (C) The typical posture for smelling a mark of a conspecific; note the open mouth, and retracted tongue.

tions) may also be deposited. These are the two typical urine marking behaviors of prosimians. In contrast, salivary marking has been observed only in *Microcebus* species (*M. coquereli* and *M. murinus*).

A peculiar trail of *M. coquereli* is the "odor discharge" mentioned above. This type of volatile odor release has also been observed in other prosimians (Schilling, 1978), but no relation with territoriality has been observed. However, Schilling (1980) has demonstrated that *M. coquereli* can discriminate at a distance from minute and dry urine traces: sex and sexual state, individual identity, and time of deposition.

Male-Female Relations

Male 6, on which most observations were made, encountered a conspecific on an average of every other night. However, prolonged contact (with ♀2) was observed less frequently. These encounters occurred either in large bare trees (Boababs or Harofy) or in bushes with dense foliage.

Encounters between ♂6 and ♀2 often included movements as a pair accompanied by constant vocal contact (contact calls), rest periods, body contact, and reciprocal grooming or social play. As mentioned above, these activities characteristic of pair bond relations, occur in the middle of the night, in an area common to the male and the female.

Social behavior during contacts was of two types: (1) brief encounters without tactile contact, accompanied with contact calls (this occurred between ♂6 and ♀2, ♀10, ♀15); and (2) prolonged encounters with body contact including urine licking by the male as well as play and grooming (Pages, 1980). The sequence can last up to 90 minutes even during the period of sexual inactivity (austral winter).

Brief encounters were observed between ♂6 and all the neighboring young and adult females, however prolonged contact only occurred with ♀2. This type of relationship was also observed between ♂13 and ♀10.

Even though these observations were made within the period of anestrus, privileged social contacts were observed, indicating pair bond relations. In captivity, these male-female contacts before sexual activity facilitate subsequent acceptance of the male by the female during estrus.

Aggressive contacts between male and female have also been observed (♀10 with ♂14 and ♂?), including chasing and fighting, sometimes resulting in both animals falling down to the ground.

Male-Male Relations

Two types of relationships could be distinguished, which do not depend on the individuals in contact, but rather on where the encounter takes place.

1. Encounters in the periphery of the home range, usually after midnight, do not involve body contact or chasing, but may be accompanied by calls which result in retreat of one of the individuals (observed between ♂6 and neighboring males).

2. Encounters in the central area usually involve chases, accompanied by intense calling (male single note call and agonistic calls). An example of this occurred when ♂6 was released after several days of captivity: on three occasions, he chased other males who had invaded the central area of his home range during his absence.

These two types of encounters again illustrate the differential use of the two areas of the home range: a peripheral area in which numerous unspecific encounters occur, and a defended central area in which food resources are obtained, and privileged male–female contacts take place.

Female-Female Relations

Encounters between females were observed on only one occasion, upon release of ♀10 who immediately began marking (urine marking and anogenital dragging) and emitting mother–infant meeting calls (one of her two young, juvenile ♂11, had not yet been released). Three neighboring females answered the mother–young meeting call (which had been abnormally repeated all night) with the female single note call. The ♀10, accompanied by juvenile ♂12, was approached by ♀15 who regularly emitted weak contact calls and attempted tactile contact. This type of female–female relation evokes that observed by Charles-Dominique (1977, 1978) between mother and daughter *Galago*.

Parent-Young Relations

In captivity, the young first leave the nest at 3 weeks of age. On the first days of exit, the young is immediately returned to the nest in the mother's mouth. The mother progressively allows longer trips out of the nest, however remains in proximity. Tactile contacts initiated by the young are avoided, and at dawn, the young is carried back to the nest or incited to follow the mother.

Gradually, at one month of age, tactile contacts are initiated by both parents and young (social play and grooming) and serve to establish stable bonds. At 40 days of age, frequent contact calls between young and adults permit auditory contact at a distance, and concurrently, exploratory behavior by the young increases. At one or two months of age, precocious sexual activity (mounting, neck biting, pelvic thrusting) by the young male and female toward the mother may be observed.

In nature, the 3-month-old juvenile of ♀ 10 (juvenile 20) was seen on different occasions either alone and searching for food, or engaged in mutual grooming and play with one or both parents (♀ 2, ♂ 6). The two young of ♀ 10 (juvenile ♂ 11 and juvenile ♂ 12, aged approximately 5 months) slept in the same nest as their mother, but normally foraged alone and exchanged contact calls. They generally left the nest after the mother and utilized only a part of the mother's home range, where she was less frequently observed.

The return to the nest, which occurred just before dawn, was accompanied by mother–young meeting calls. These mutual vocal interractions and movements allowed the mother to find the young at the end of the night. Neither the female nor the infants were ever observed to sleep with the male during the austral winter.

Discussion

The social system of *M. coquereli* is characterized by a well-defined central area, and a large variable peripheral area. In the forest of Marosalaza, in proximity of the temporary pond, where food resources (insects) are abundant, the central areas of males do not overlap and are separated by a space in which are intercalated the female/young central area (generally larger). The female central area may partly overlap with that of a male, depending on pair bond relations.

This pattern of spatial distribution forms a mosaic which efficiently covers the available space and assures maximum utilization of food resources. If the role of the central area is to provide a sufficient food supply, the peripheral area permits social exchanges.

This spatial separation is paralleled by a structured temporal distribution of the different types of activities during the night (Fig. 9). The regular occurrence of solitary activities and social encounters at different times probably allows the species to maximally exploit food resources of the central area at the beginning of the night: the first part of activity is mainly devoted to feeding, so that on cold nights, when there is a reduction in total activity, this mainly affects the frequency of social encounters.

During winter, in captivity, feeding activity was more regularly distributed throughout the night (see Chapter 7, and Pages, 1980). However, this was probably due to the constant availability of food, and to the reduced space which did not allow spatial distribution of activities. In contrast, the circadian rhythms of social activities closely approximated that observed in nature.

During summer, there is a general increase of activity, and particularly

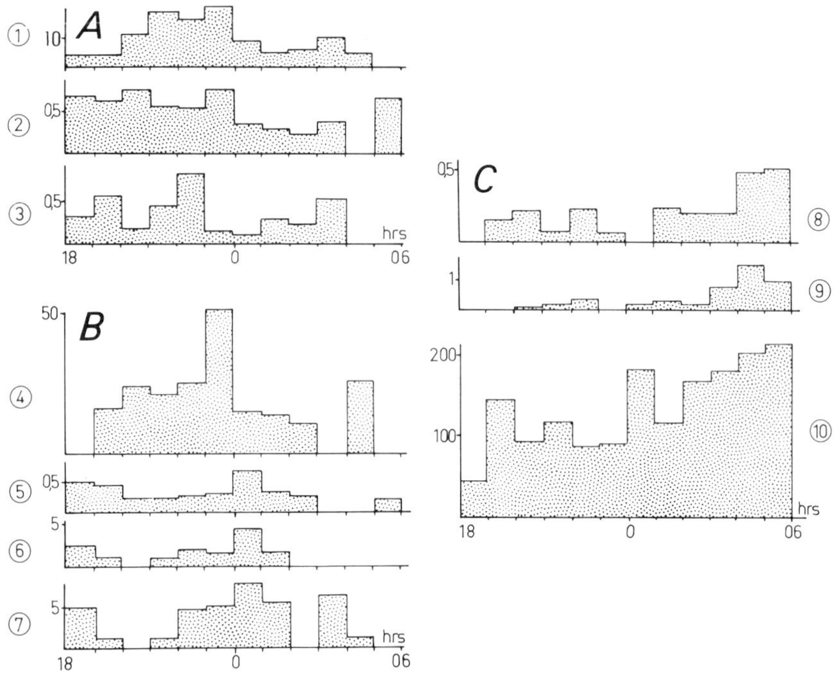

Fig. 9. Activity rhythms for over 100 hours of observation of Male 6: (A) Solitary and feeding activities, which are most frequent during the first half of the night: 1. Rest and self grooming, in minutes per hour. 2. Average number of homopteran colonies visited. 3. Average number of times insects were chased and eaten. (B) Activities which are most frequent in the middle of the night: 4. Percentage of time spent at height above 15 m (presumably related to territorial defense). 5. Average number of scent release. 6. Social play with Female 2, in minutes per hour. 7. Prolonged contact with Female 2, in minutes per hour. (C) Activities which are most frequent at the end of the night. 8. Average number of encounters with conspecifics other than Female 2. 9. Average number of vocalizations. 10. Average distance (in meters) covered during different hours of the night.

social activities. At the time of estrus, social behavior becomes increasingly important, and often occurs at the beginning of the night and may replace a large part of feeding activity. This situation may reflect that occurring in nature during austral summer, when resources are abundant enough to require only a small part of the time budget to be collected.

During the season of sexual inactivity, it appears that a minimum of social contacts are necessary to ensure successful reproduction during the following season. The maintenance of privileged relations with the parents and between young of the same litter are beneficial to the young in the development of feeding and social behaviors. The establishment of pair relations before the period of sexual activity has been observed to be favorable to reproduction in captivity by reducing the female's aggressivity towards the male.

The particular cycle of activity of *M. coquereli* is also a complement of the different ecological and behavioral characters permitting niche separation among the five nocturnal prosimian species inhabiting the Marosalaza forest. Among them, only *Lepilemur ruficaudatus* does not include insects in the diet. The three other species, *Microcebus murinus, Cheirogaleus medius,* and *Phaner furcifer* are potential competitors for animal food (Chapter 2). Together with the specialization on insect secretion, providing a regular food supply during the dry season, the temporal patterning of activities of *Microcebus coquereli* (which feeds at times different from the other prosimians), tends to reduce interspecific competition.

REFERENCES

Charles-Dominique, P. (1977). "Ecology and Behaviour of Nocturnal Primates." Duckworth, London.

Charles-Dominique, P. (1978). Solitary and gregarious prosimians: Evolution of social structures in Primates. *In* "Recent Advances in Primatology" (D. J. Chivers and K. A. Joysey, eds.), Vol. 3, pp. 139–149. Academic Press, New York.

Charles-Dominique, P., and Bearder, S. K. (1978). Field studies of Lorisid behavior. *In* "The Study of Prosimian Behavior" (G. A. Doyle and R. D. Martin, eds.), pp. 567–629. Academic Press, New York.

Hladik, C. M. (1978). Diet and ecology of Prosimians. *In* "The Study of Prosimian Behavior" (G. A. Doyle and R. D. Martin, eds.), pp. 307–357. Academic Press, New York.

Pages, E. (1978). Home range, behaviour and tactile communication in a nocturnal malagasy lemur, *Microcebus coquereli. In* "Recent Advances in Primatology" (D. J. Chivers and K. A. Joysey, eds.), Vol. 3, pp. 172–177. Academic Press, New York.

Pages, E. (1980). Relations sociales et importance des échanges tactiles dans la cohésion familiale chez *Microcebus coquereli,* lémurien malgache primitif. *Acad. Malgache* (in press).

Pariente, G. F. (1974). Influence of light on activity rhythms of two malagasy lemurs *Phaner furcifer* and *Lepilemur mustelinus leucopus. In* "Prosimian Biology" (R. D. Martin, G. A. Doyle, and A. C. Walker, eds.), pp. 183–198. Duckworth, London.

Petter, J. J. (1978). Ecological and physiological adaptations of five sympatric nocturnal lemurs to seasonal variations in food production. *In* "Recent Advances in Primatology" (D. J. Chivers and J. Herbert, eds.), Vol. 1, pp. 211–223. Academic Press, New York.

Petter, J. J., and Charles-Dominique, P. (1978). Vocal communication in Prosimians. *In* "The Study of Prosimian Behavior" (G.A. Doyle and R.D. Martin, Eds.), pp. 247–305. Academic Press, New York.

Petter, J. J., Schilling, A., and Pariente, G. F. (1971). Observations éco-éthologique sur deux Lémuriens nocturnes *Phaner furcifer* et *Microcebus coquereli. Terre Vie* **3**, 287–327.

Petter, J. J., Albignac, R., and Rumpler, Y. (1977). Mammifères Lémuriens (Primates prosimiens). *Faune Madagascar* **44.**

Petter-Rousseaux, A. (1968). Cycles génitaux saisonniers des lémuriens malgaches. *Entretiens Chize,* **1,** *Ser. Physiol.* 11–12.

Petter-Rousseaux, A. (1974). Photoperiod, sexual activity and body weight variations of *Microcebus murinus*. *In* "Prosimian Biology" (R. D. Martin, G. A. Doyle, and A. C. Walker, eds.), pp. 365–373. Duckworth, London.

Schilling, A. (1978). Olfactory communication in Prosimians. *In* "The Study of Prosimian Behavior" (G. A. Doyle and R. D. Martin, eds.), pp. 461–542. Academic Press, New York.

Schilling, A. (1980). The Possible Role of Urine in Territoriality of Some Nocturnal Prosimians. *Symp. Zool. Soc. London* (in press).

Zimmermann, F., Gerard, H., and Charles-Dominique, P. (1976). Le radio-tracking des Vertébrés: Conseils et techniques d'utilisation. *Terre Vie* **30**, No. 3, 309–346.

5 Quantitative and Qualitative Study of the Light Available in the Natural Biotope of Malagasy Prosimians

G. F. PARIENTE

INTRODUCTION

Nocturnal vision poses numerous problems, particularly concerning animals which use the eye as the main sensory system in very low light levels. Whether one considers birds, fruit-bats, or nonflying mammals, efficient solutions have been adopted during the course of evolution to permit sufficiently "comfortable" vision for the detection of objects in space. To speak of an efficient solution implies a problem, in this case essentially that of the quantity of light available in conditions where the species normally live.

In many diurnal animals the ability to see at night is weak or absent (such is the case for "civilized" man) and those species which are "at ease" during the night have often acquired several anatomical and/or physiological specializations (*tapetum lucidum,* photoreceptor convergence) which strongly suggests that, without the advantage thus acquired, the quantity of nocturnal light would be insufficient for effective vision. These various adaptations have already been studied by many authors for nocturnal birds of prey, fruit-bats, and some species of carnivores (Tansley, 1965).

Concerning the vision of prosimians (see Pariente, 1978, for review), few studies have been published and no efforts have been directed toward an understanding of the spectral environmental conditions in which these species are normally active. In this chapter, results are presented on the quantity and quality of light available, in different forests of Madagascar, and particularly in Marosalaza, where the observations presented in the other chapters of this volume have been conducted.

METHODS

Material

The remoteness of the chosen study areas required utilization of a light detection system which was transportable and autonomous, but also capable of functioning under extremely low light levels. A microamperemeter (''International Light Society'' IL 600) adaptable to several captors and a photomultiplier system were chosen for this study. The limits of the spectral zone studied were defined by each of the available captors:

1. daytime measures

PT 100 permitted measures between 200 and 650 nanometers (nm)
PT 200 A permitted measures between 450 and 1000 nm

2. nightime measures

The combination of the IL 600–IL 660 with the photomultiplier PM 200 C, covered a band between 200 and 650 nm.

At the time of this study, these were the only types of captors available which fulfilled our requirements of high sensitivity plus autonomy.

In practice, the response curve of these three captors could be slightly extended (100 nm) into a zone of poor sensitivity which made possible, in certain instances, an extension of the range of measures. The additional use of filters (12 interference, 4 optical) permitted measures in limited and well-defined bands of the spectrum.

Method of Energy Calculation

The calibrated curve of relative sensitivity of each captor gives the response to 1 μW at intervals of 10 nm. The values thus obtained yield a curve of relative sensitivity of the apparatus as a function of wavelength.

In addition, for each captor a reference point was measured at absolute values:

with the PT 100 at 100 nm: 1 $\mu W/cm^2$ yielded 1.522×10^{-2} μA
with the PT 200 at 800 nm: 1 $\mu W/cm^2$ yielded 5.05×10^{-3} μA
with the PM 200C at 400 nm: 1 $\mu W/cm^2$ yielded 3.843×10^{4} μA.

These three values permitted subsequent calculation of absolute values from the curves of relative sensitivity throughout the spectral range.

The use of filters required point by point consideration of the response of the captors with each filter. The transmission curve of each filter was plotted as a function of wavelength (Laboratoire National d'Essais). Since the filters had unequal transmission factors, all estimations were standardized to a calculation of the energy contained in a band 1 nm wide, in order to render measures comparable.

For simplification (due to the impossibility of using more narrow band filters because of the low light levels measured), it was assumed that for each combination captor/filter, the source was uniform in the corresponding band of transmission. The maximum transmission of each filter was used in the curves for the calculation of the energy contained in a band 1 nm wide.

Thus, for each combination captor/filter, a coefficient was established permitting direct calculation of the energy, at a given wavelength as defined above, considering the assumption concerning emission of the light source.

In the cases where the large band optical filters were used (or no filter at all), it was assumed that the source was "solar" light, which was defined for three color temperatures (5000°K, 6500°K, 7500°K).

An additional calculation permitted comparison of the sensitivity of our apparatus to that of the eye of a human observer. The combination captor/filter/solar source at the chosen temperature for the standard observer (normal CIE scotopic or photopic curve) yielded conversion coefficients of the measured current into lux (Le Grand, 1972). It should be evident, however, that the absolute values of energy calculated in this study are not assumed to be very precise. In fact, it is the relative values which are the most interesting to consider. At low light levels errors due to reading of the meter and fluctuations in the power source, may lead to systematic error of $\pm 20\%$. Continuous recordings in these same conditions result in irregular curves mainly due to different meteorological phenomena. Therefore, the measures yielded an estimation of absolute values, and the possibility of comparing relative values.

Finally, after comparison of the curves (by direct measure of solar energy) with classic curves of the solar spectrum, we always aligned the PT 200 A with the PT 100 by making an adjustment at 550 nm (the two captors do not give exactly the same values at the same wavelength).

Standardization of Measures

In order to proceed under comparable conditions the captors were placed at one meter from the ground and oriented towards geographical North at each site (the magnetic variation is very marked in certain regions of Madagascar, and reaches values of 18–19°W near Morondava).

At the beginning of this study, measures were taken every 2 hours and in five directions (North, South, East, West, and Zenith for each filter). Ground illumination was occasionally measured. In later work, in order to reduce the amount of time between measures (20 minutes in the above method), records were made toward the zenith (↑) and toward the South (S).

In the latter case this avoids direct light which is greater toward the North in the Southern hemisphere. Measures of the sun and the moon permitted calculation of the energy spectrum of these two main light sources. With these methods the time required to complete a series of 12 measures was reduced to 7 minutes.

RESULTS

Measures outside the Forest at Marosalaza (50 km North of Morondava)

Study of the Solar Source

These preliminary measures were taken to verify the proper functioning of the apparatus. Figure 1A and B shows the results obtained at different times of the day with clear weather conditions, and illustrates the well-known effect of color temperature variation of the solar source as a function of the incident angle of light rays through the atmospheric layers: red is dominant at 7.00 hours, and a strong relative increase in the short wavelengths occurs before noon (Fig. 1A). Measures made before sunset under the same conditions are illustrated in Fig. 1B. It should be noticed that measures taken either directly toward the sun or in other directions yield about the same relative distribution of energies according to wavelength. As an example, Fig. 2 shows measures taken toward the south (dominance of long wavelength) which should be compared to Fig. 1.

Study of the Nocturnal Sky

The changes in light radiation outside the forest at dusk can be seen in Fig. 3A. At 19.00 hours, it was not possible to use the PM at all wavelengths due to the relatively intense light levels. The energy distribution during full nighttime (Fig. 3B) shows a continuous relative increase in energy toward the longer wavelengths. At wavelengths of 800 nm and longer, the level of energy is too weak to be detectable by the PM with the interference filters used in this study. Figure 3B also shows that there is little variation in the relative distribution of energy between the zenithal (↑) or southern (S) direction. However, there is a difference in intensity, zenithal light being significantly higher. These latter curves will be compared (see

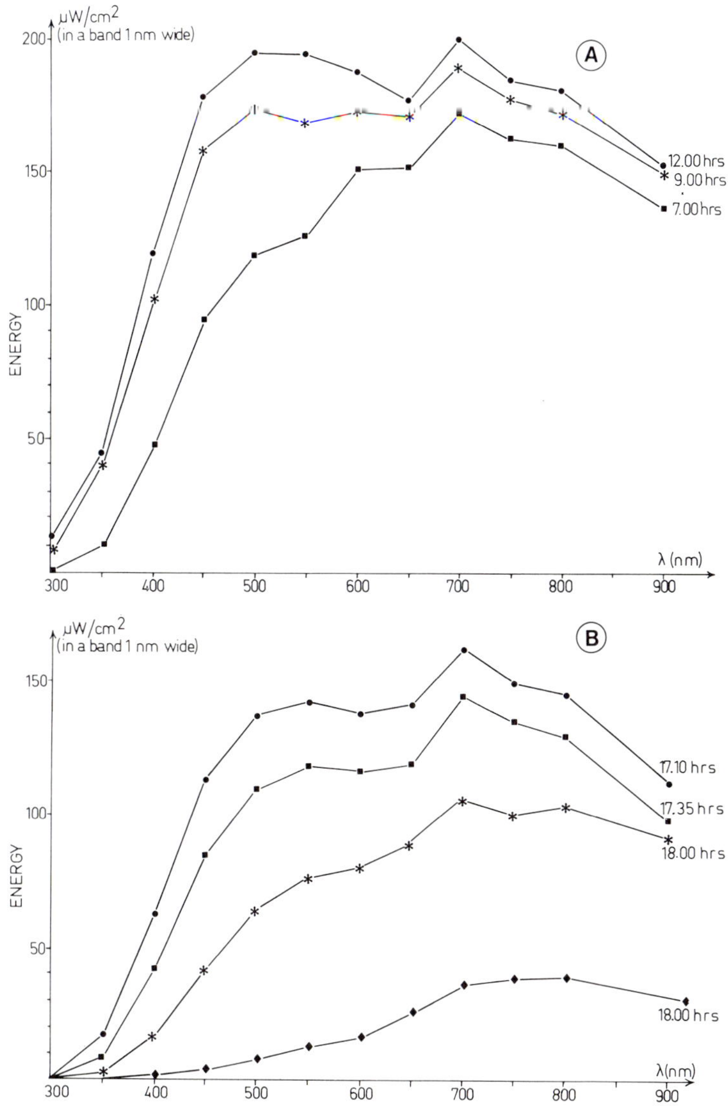

Fig. 1. Distribution of solar energy as a function of wavelength at different times of the day outside the Marosalaza forest. (A) After sunrise (5:58 hours, February 27, 1974); (B) Before sunset (18:30 hours, January 14, 1974).

below) to those obtained in the forest under the same conditions since they most closely approximate the actual energies available to the eye of an animal which is moving along a basically horizontal plane (we assume that in general the direction of gaze is horizontal and permits image formation

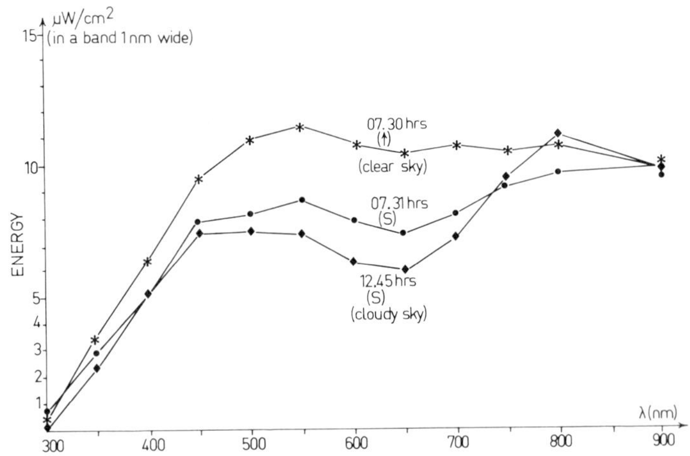

Fig. 2. Distribution of spectral energy of the daylight. Measures taken toward the zenith (↑) and the south (S), outside the Marosalaza forest (February 27, 1974).

grossly corresponding to analysis of the vertical supports of the environment).

Importance of the Moon

The effect of the lunar cycle (hours of moonrise and moonset) and the influence of the lunar phase on the luminous environment have already been discussed elsewhere (Pariente, 1976, 1978). The present discussion will be limited to consideration of the moon as a light source, under the best possible meteorological conditions (clear sky, moon near the meridian), as has been done above for the sun.

In Fig. 4 are presented values taken by direct measurement of the moon, either at the time of full moon or during the first quarter. The light reflected by the moon is primarily dominated by the longer wavelengths. It may also be seen that the ratio of intensity between the full moon and the first quarter is in the order of 1:10 and, during the full moon there is a marked qualitative variation typified by a relative decrease in energies beyond 750 nm. At the first quarter (the light rays are subjected to reflection at an angle of 45° on the moon) the light is rich in radiations of longer wavelength. It thus may be assumed that for a nocturnal prosimian eye having a maximum of sensitivity near 500 nm (Pariente, 1976, 1978; Alfieri *et al.*, 1974) the difference in the relative spectral distribution between first quarter and full moon is negligible.

A comparison of the curves of Fig. 1A and B and Fig. 4, at 500 nm, shows that the moon yields only one millionth the energy provided by the sun at noon.

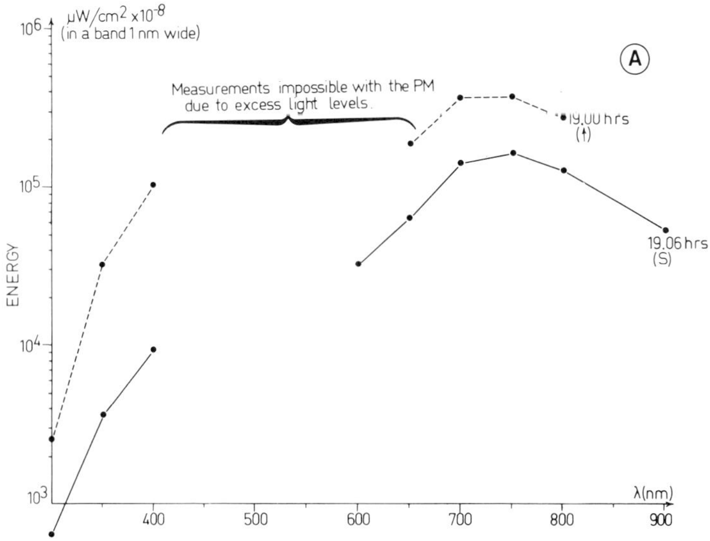

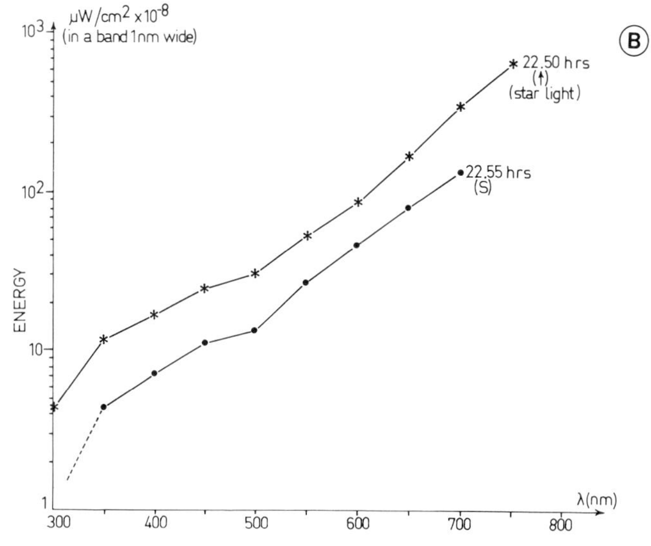

Fig. 3. Distribution of spectral energy of the nocturnal light. Measures taken toward the zenith (↑), and toward the south (S), outside the Marosalaza forest. (A) At dusk (Jan. 14, 1974); (B) during the night (Feb. 23, 1974).

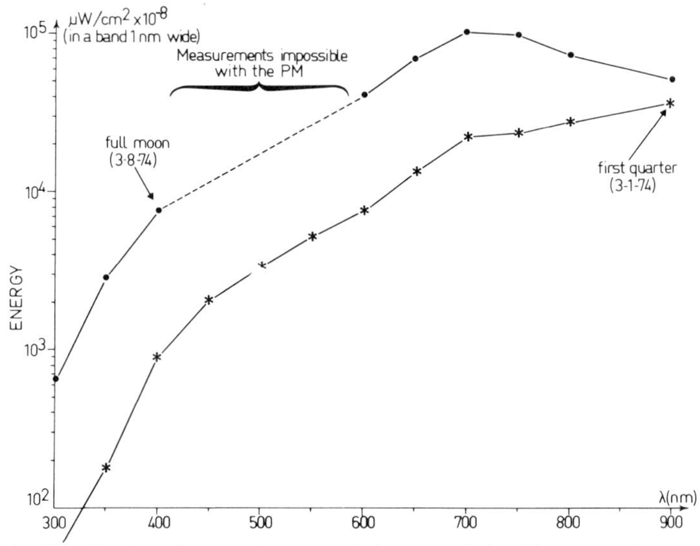

Fig. 4. Distribution of spectral energy of the moon light. Measures taken outside the Marosalaza forest (March 1, 1974).

Study of Light in the Forest at Marosalaza

Since direct measures of the sun or the moon were not possible in the forest, only measures toward the south and zenith were performed.

Forest Daylight

Since the study period took place during the rainy season (austral summer) the sky was not always perfectly clear. However, the curves presented in Fig. 5 show the distribution of radiation when the sky is only slightly covered. This curve should be compared to that of Fig. 1 (sun source) and to Fig. 2, which gives an idea of the screening effect of vegetation in Marosalaza. It can be seen that toward the zenith the curves are very similar, the vegetation slightly modifies the light in intensity but not in quality. This result is not surprising since in the forest of Marosalaza the vegetation is relatively sparse and direct rays from the sun easily reach the ground through the foliage, especially during the period between 10.00 and 14.00 hours.

The intensity of light from the south, which mainly determines the actual visual conditions in the undergrowth, is about ten times less than that from the zenith. Therefore, at 500 nm the effect of the vegetal cover reduces the light level by a factor of 40 compared to light conditions outside the forest.

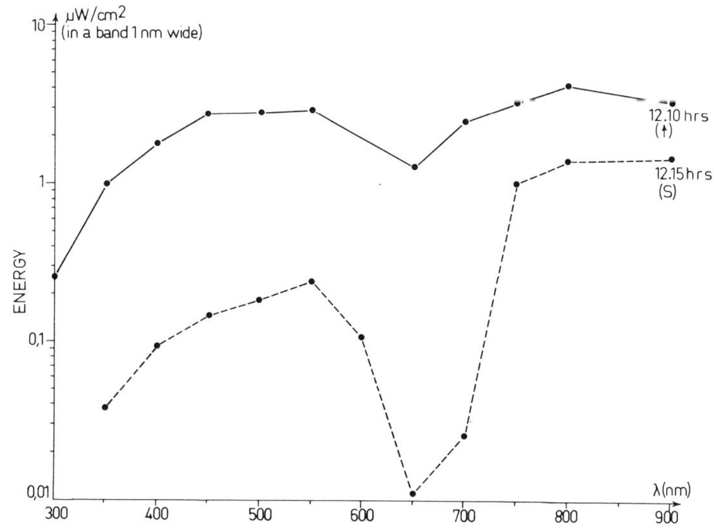

Fig. 5. Distribution of spectral energy of light inside the Marosalaza forest during daytime. Measures taken toward the zenith (↑) and the south (S).

There is especially a strong absorption of radiation between 500 and 750 nm, which corresponds to absorption by chlorophyll and other pigments. However, while chlorophyll also strongly absorbs in the blue spectral zone, this is not very apparent in Fig. 5.

Forest Nocturnal Light

Due to the extremely low energies at night, measures were only possible in the band of maximum sensitivity of the PM (between 200 and 650 nm), particularly in the southern direction. In Fig. 6 the results are presented for measures taken toward the south at different times of night. Depending on the time of measure the spectral energy distribution varies greatly. It appears that after 19.28 hours (when there is still a slight absorption between 500 and 650 nm, probably due to vegetal pigments) the curves tend toward a more even distribution increasing from blue to red. At 0:45 hours a strong blue contribution may be noticed, which may be due either to the appearance of a brilliant constellation or to a clearing of the night sky. Indeed, the origin of light at night is very varied and includes stellar light (30%), zodiac light (30%), and aurora australis, diffusion of solar light by the atmosphere, photo-emission from high energy particles and various other sources.

In order to appreciate these phenomena we have plotted in Fig. 7 the

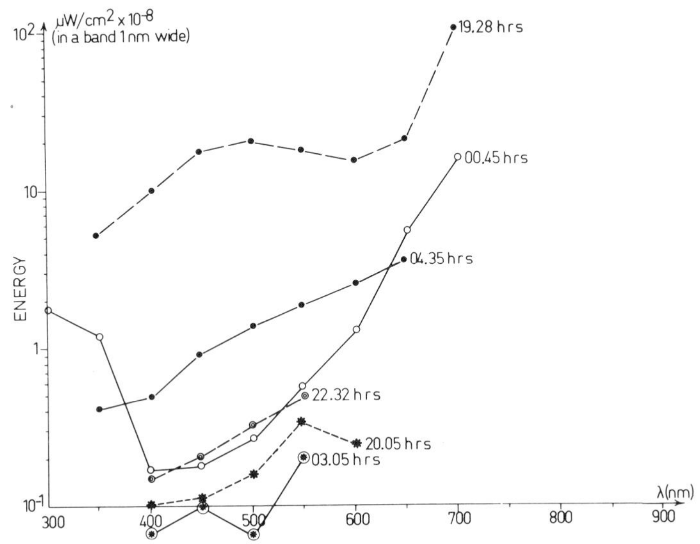

Fig. 6. Distribution of spectral energy of the light inside the Marosalaza forest, at different times of the night (January 18/19, 1974). Measures taken toward the south.

variation in energy according to time of night at two wavelengths, 450 and 499 nm (southern direction). At 00.45 hours, when the brilliance of the stars appeared exceptional, there is a slight increase in energy. On the other hand, the low values after dark and just before dawn may be due to slight meteorological effects such as condensation due to a decrease in temperature. This type of variation, including a slight increase around midnight was often observed. Thus, many parameters can intervene to modify the spectral energy distribution. The marked increase in radiation at 499 nm at 04:35 hours is due to the presence of the moon (rise at 02:05 hours, age = 25 days, a fine crescent at 30° from the horizon). A weak effect is only noticeable 2 hours after moonrise, especially in the longer wavelengths. However, it cannot be excluded that these changes may also be due to the first effects of sunrise which on this day (January 19, 1974) occurred at 05:37 hours, one hour after the measures were taken. In relation to this increase in light intensity at 04:35 hours there is a sudden increase in activity of *Phaner furcifer* (Pariente, 1974, 1978).

It is of course useful to give an idea of what these values represent for a human observer (conditions of scotopic vision). The values obtained at 00:20 hours on the night of January 18–19, 1974, in the Marosalaza forest during a clear night without the moon, corresponded to 5.3 μlux in a southern direction, and 134 μlux toward the zenith. The luminance of the sky seen through the foliage equaled 3.6 × 10^{-5} μW/cm² between 200 and

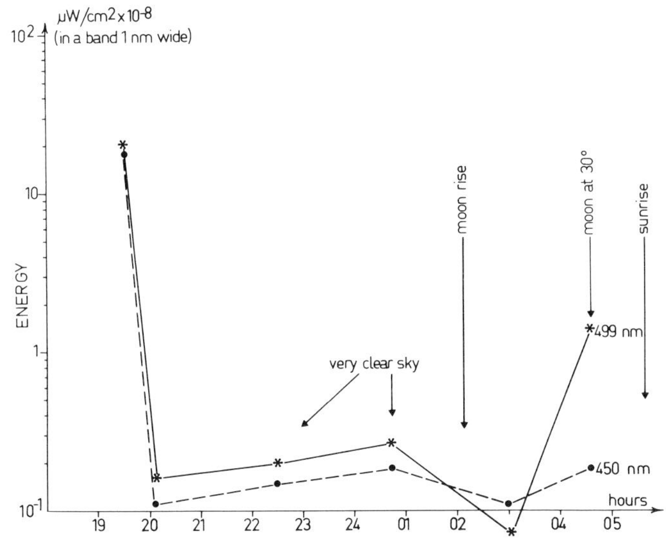

Fig. 7. Variation in energy at 450 nm (dashed line) and 499 nm (solid line), at different times of the night. Measures taken toward the south inside the Marosalaza forest (January 18/19, 1974).

650 nm. Under these conditions it was totally impossible to distinguish the forest environment. At 19:30 hours, when 160 lux were available from the southern direction, our vision already started to deteriorate; however, a few minutes before at 19:23 hours the zenith still produced 11 millilux and the south 350 μlux. At this moment we could still just perceive the general outline of the larger trees in the immediate vicinity. This simple example illustrates how merely a factor of 30 can alter an observer's visual performance from a threshold level, still permitting useful vision, to the total impossibility of seeing. This demonstrates that the eye of a strictly nocturnal animal only needs relatively slight perfectioning in order to function efficiently.

Comparative Measures Made in Madagascar at Other Locations

These measures, performed in 1970, give a global representation of the light in different biotopes where prosimians are found but are less complete due to the lack of interference filters. The methods were identical to those given above, except that only wide band, or no filters were used.

Eight locations in Madagascar were studied:

Fort-Dauphin, Southeast

Berenty (gallery forest), South
Tulear (bush), Southeast
Analabe (Marosalaza) (dry deciduous forest), West
Ambohimanarina (central plateau)
Perinet (humid evergreen forest), East
Ankarafantsika (deciduous forest), Northeast
Diego-Suarez (humid evergreen forest), North

The results presented in Fig. 8 require some qualifications:

In several cases, circumstances did not allow all the measures to be made, thus some graphs are incomplete.

At Tulear, the measures were made in sparse "bush," comparable to open territory.

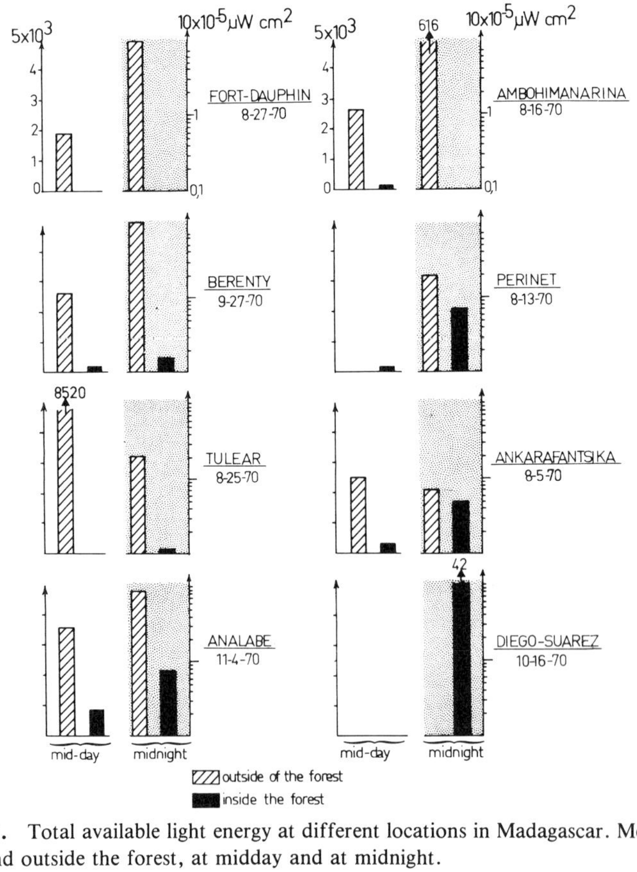

Fig. 8. Total available light energy at different locations in Madagascar. Measures taken inside and outside the forest, at midday and at midnight.

In two cases, the presence of the moon resulted in higher values (Ambohimanarina and Diego-Suarez).

An interval of 3 months elapsed between the first and last measures.

Measures Made during the Day

The daytime values at Tulear are quite exceptional. This is probably due to the proximity of the Mozambique channel which is responsible for a significant diffusing fog (which affects all visible radiations). As a result the luminosity is extremely intense. The same type of situation was also found at Morondava and Analabe (Marosalaza forest) during the rainy season.

In the different forests variation is low, and depends upon the type of vegetation (large trees at Berenty, canopy of average thickness at Analabe (Marosalaza), sparse forest at Ankarafantsika, and relatively clear forest on the slopes and crests where we worked in Perinet.

Measures at Night

At night these values are similar outside the forest for all the stations, except at Tulear where the values are lower, probably due to the high humidity (which in this case produces an effect opposite to that observed in sunlight).

Energy Distribution

The use of large band optical filters, some of which transmit elsewhere than in the zone considered, did not permit precise quantitative calculations. However, the combination of curves of the PM 200C and PT 100 with filters of 350 nm and 640 nm (peak of transmission) permitted a "relative" estimate of the ratio of blue and red wavelengths (Fig. 9). In full daylight, outside the forest, the ratio of blue to red is in all cases inverse to that at night. However, if measures made outside the forest are compared to those made inside the forest, it may be noted that there is generally an inversion of the proportions of blue and red (see Berenty open and Berenty forest). Inside the Marosalaza forest, the values of blue and red are not significantly different (measures made during austral winter, the dry season). In two cases (Ambohimanarina and Diego-Suarez) the presence of the moon accounted for the exceptionally high values.

DISCUSSION

The results presented in this study give an idea of light conditions which may influence the visual processes of the Prosimians. The main findings of interest are the following:

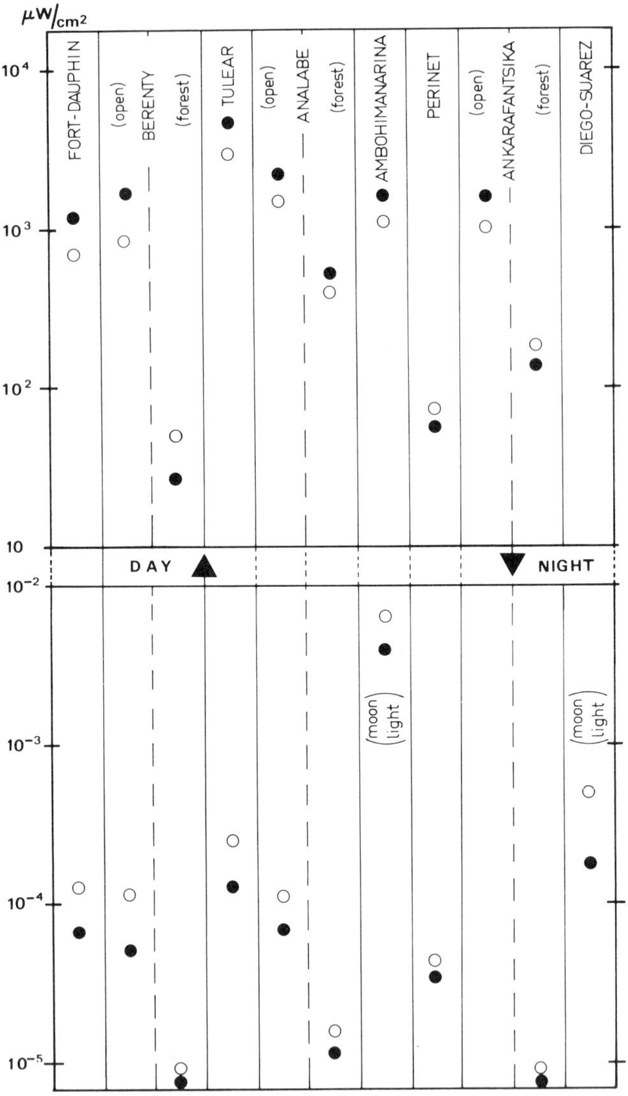

Fig. 9. Comparison of light available during the daytime and at night, in either dominant red (open circles) or dominant blue wavelength (dark circles) at different locations in Madagascar, inside and outside the forest.

Outside the forest, there is a relative homogeneity in daylight levels except in the southwestern coastal region where meteorological phenomena add to direct radiation.

Inside the forest during the dry season, daylight is only moderately reduced, generally by a factor of ten from the south, depending on vegetation structure.

During the rainy season, when vegetation is dense, this coefficient of absorption may increase considerably, and reach values up a factor of sixty.

In the forest at night, the absorptive effect of the vegetation on incident light is greatly increased.

Color temperature is generally high during the day and rapidly decreases with decreasing light conditions. The night is thus characterized by a very marked "reddish quality."

During the periods of dawn and dusk, very rapid variations occur in intensity and spectral composition.

In addition, at dawn (Fig. 10) there is a rapid increase in intensity, which differs according to the direction of measure. Light from the zenith,

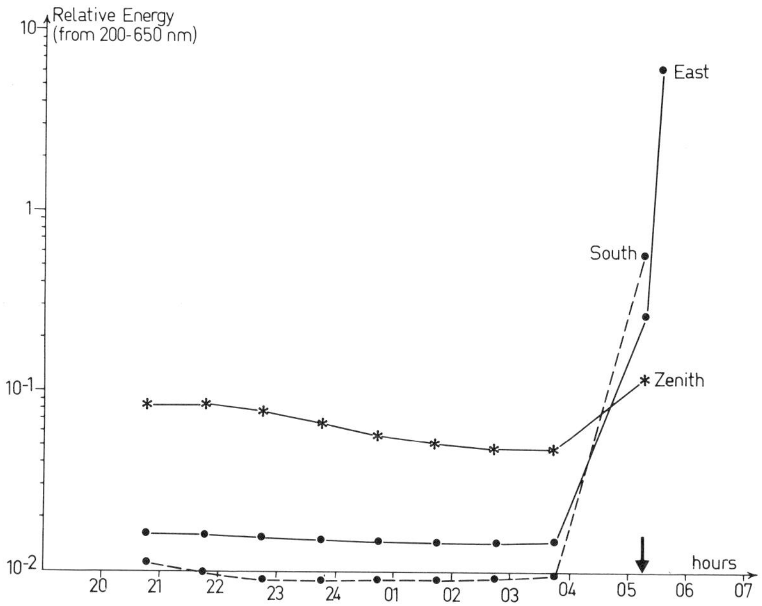

Fig. 10. Distribution of energy between 200 and 650 nm at night inside the forest (Ampijora, August 4, 1970). Measures taken in different directions. The arrow indicates the time of minimum intensity necessary for human vision.

which is initially the most intense due to the presence of stars, becomes relatively reduced compared to other directions. Light from the east increases most rapidly due to light rays diffracted by the atmospheric layers.

These measures will provide a background for a better understanding of the visual limits of Prosimians, and the relation to behaviors, such as activity rhythms (Pariente, 1974, 1975b; Chapter 7), and visual communication (Pariente, 1978). Hopefully, this work will aid both ethologists and ecologists studying the impact of "light phenomenon" on species evolution.

NOTE

The relationships between the light environment and its influence on the visual system and behavior was always of primary importance in G. Pariente's research. The scope of his studies was large and dealt with the influence of light on activity rhythms (1974, 1975a,b), the role of light in visual communication (1978), anatomical studies of the eye (1970, 1975a,b, 1976), and behavioral and electrophysiological studies of relative luminous sensitivity (1974, 1976).

To illustrate the correlations which may be established between quan-

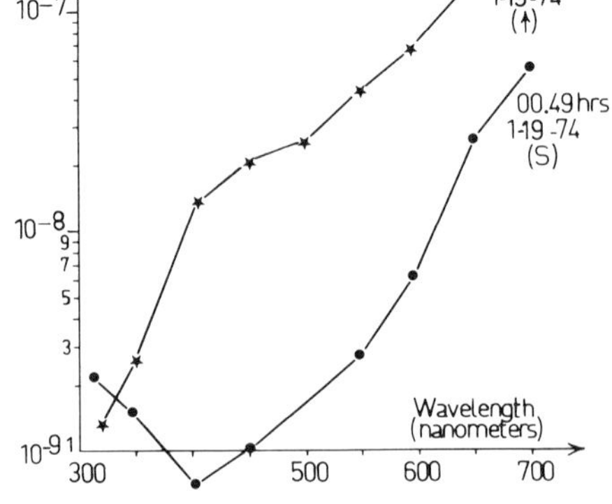

Fig. 11. Distribution of spectral energy at night in the Marosalaza forest. Comparison of measures taken toward the zenith (↑) and the south (S) (data from Pariente, 1976).

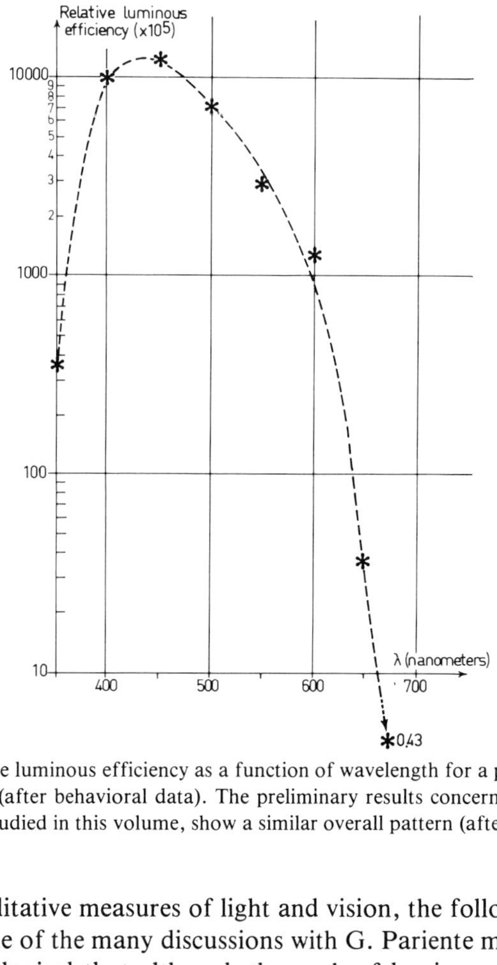

Fig. 12. Relative luminous efficiency as a function of wavelength for a prosimian species, *Hapalemur griseus* (after behavioral data). The preliminary results concerning certain other prosimian species studied in this volume, show a similar overall pattern (after Pariente, 1976).

titative and qualitative measures of light and vision, the following example derived from one of the many discussions with G. Pariente may be cited. It may seem paradoxical that although the peak of luminous efficiency for nocturnal prosimians occurs in the blue-green part of the spectrum (450–500 nm), the reflected nocturnal light in the forest is mainly dominated by the longer wavelengths.

In fact the explanation for this discrepancy becomes apparent when the total light environment is considered. Horizontal reflected light in the forest is relatively poor in blue-green wavelengths (due to the absorption); whereas light from the zenith is relatively rich in blue-green wavelengths. The resulting difference in energy between the "reddish" forest light and direct "bluish" light is maximum at 450–500 nm (Fig. 11). This is exactly in the region of peak sensitivity for nocturnal prosimians (Fig. 12). Thus, the vision of these nocturnal species seems to be adapted to utilize contrast cues for detecting objects directly illuminated by the "bluish" zenithal light source against the dark "reddish" forest background.

REFERENCES

Alfieri, R., Pariente, G. F., and Sole, P. (1974). Dynamic electroretinography in monochromatic lights and fluorescence electroretinography in Lemurs. *Int. Soc. Clin. ERG Symp., 12th, 1974.*

Le Grand, Y. (1972). Spectral luminosity. *In* "Handbook of Sensory Physiology" (D. Jameson and L. M. Hurvich, eds.), Vol. 7, Part 4, pp. 413–433. Springer-Verlag, Berlin.

Pariente, G. F. (1970). Rétinographies comparées des Lémuriens malgaches. *C. R. Hebd. Seances Acad. Sci.* **270,** 1404–1407.

Pariente, G. F. (1974). Influence of light in the activity rhythms of two malagasy lemurs: *Phaner furcifer* and *Lepilemur mustelinus leucopus. In* "Prosimian Biology" (R. D. Martin and G. A. Doyle, eds.), pp. 183–198. Duckworth, London.

Pariente, G. F. (1975a). Observations ophtalmologiques des zones fovéales vraies chez *Lemur catta* et *Hapalemur griseus* (Primates de Madagascar). *Mammalia* **39,** 143–151.

Pariente, G. F. (1975b). Lumière et rythme d'activité de *Phaner furcifer* (Prosimien nocturne de Madagascar) dans son milieu naturel. *J. Physiol (Paris)* **70,** 637–647.

Pariente, G. F. (1976). Etude éco-physiologique de la Vision chez les Prosimiens malgaches. Thèse de Doctorat d'Etat, Montpellier.

Pariente, G. F. (1978). The role of vision in Prosimian behavior. *In* "The Study of Prosimian Behavior" (G. A. Doyle and R. D. Martin, eds.), pp. 411–459. Academic Press, New York.

Tansley, K. (1965). "Vision in Vertebrates." Cox & Wyman, London.

PART II

Studies in Simulated Climatic Conditions

6 Seasonal Activity Rhythms, Reproduction, and Body Weight Variations in Five Sympatric Nocturnal Prosimians, in Simulated Light and Climatic Conditions

A. PETTER-ROUSSEAUX

INTRODUCTION

Previous studies concerning the nocturnal prosimian *Microcebus murinus* have shown that seasonal variations exist in body weight, temperature, and sexual activity which depend on variations of the photoperiod (Petter-Rousseaux, 1970, 1974). In the present chapter, observations have been extended to include five species of sympatric nocturnal prosimians, and cover a period of 3 years. Data were collected on body weight variations, food intake, locomotor activity, and reproduction, to allow comparison of the specific physiological cycles. A more detailed analysis of circadian rhythms is reported in Chapter 7.

These species (*Microcebus murinus, Microcebus coquereli, Cheirogaleus medius, Phaner furcifer,* and *Lepilemur ruficaudatus*) inhabit the Marosalaza forest (West Coast of Madagascar) where important seasonal variations occur (Chapter 1). During field work covering almost an entire year, these species were the subject of different studies presented in other chapters of this volume. The results of these field studies raised certain questions concerning specific physiological adaptations which were difficult to analyze in natural conditions. In particular, *Cheirogaleus medius* has been observed to hibernate for a period of several months during the austral winter, and *Microcebus murinus* undergoes short periods of "lethargy" (Petter, 1978; Chapter 2).

It was thus of interest to raise these species in captivity, in conditions resembling as closely as possible those of the natural environment. Thus we attempted to reproduce in the animal house of Brunoy, the different parameters actually recorded in the Marosalaza forest (Chapters 1 and 5).

METHODS

Housing

The animals were housed in a room measuring 8 × 5 × 4 m, divided by wood and glass walls into 6–8 m^2 cubicles, the roof of which consisted of wire mesh at a height of 2 m. Nest boxes and branches were provided in each of these cages.

Each cubicle contained individuals of only one species:

Microcebus murinus (body weight approximately 80 gm), one male and four females

Microcebus coquereli (body weight approximately 350 gm), one male and one female plus a young male born in July 1973

Cheirogaleus medius (body weight approximately 300 gm), one male and one female

Phaner furcifer (body weight approximately 450 gm), one male and two females

Lepilemur ruficaudatus (body weight approximately 700 gm), two males from March 1973 until June 1974; two females added in June and September 1974. Couples were housed in separated cages in September 1974.

Lighting System

Two independent systems were used to control the light cycle: (1) three sets of five fluorescent tubes on the ceiling, controlled by a timer circuit; and (2) two sets of six tubes for progressive increase/decrease in light intensity, located at each extremity of the room, controlled by a timer circuit which simulated twilight lasting approximately ½ hour.

During artificial night periods, red lights permitted visibility of the interior of the cubicles.

The daily variations in light closely approximated those of Madagascar. We chose to assign the longest days (i.e., those of austral summer) to the summer months of the northern hemisphere, in order to facilitate temperature regulation. This required the animals, most of which were recently imported at the beginning of the experiment, to readjust their seasonal physiological cycles during the first months, as will be seen.

Since all the animals were nocturnal, the periods of obscurity were arranged so that daily care of the animals took place at the beginning of the working day, before behavioral observations.

Temperature and Humidity

Heating was provided by a central heating system. The cooling of the room during the dark period was accomplished by a thermostat-controlled ventilator yielding a daily variation of 5–8°C which is less than that found in the natural environment (10–15°C). Temperatures generally varied between 24° and 30°C during the periods of long daylength (compared to 21° and 35°C in the Marosalaza forest, austral summer) and between 17° and 30°C during the periods of short daylength (compared to 12° and 30°C in the Marosalaza forest, austral winter, see Chapter 1).

No systematic effort was attempted to regulate humidity which remained comparatively low: 25–70% during the period of long daylength, and 10–40% during that of short daylength. In natural conditions, the variation is between 40 and 100% during the austral summer.

Feeding

The animals were fed mainly fruits: apples, pears, bananas, oranges, various fruits during the summer, lettuce, and the G3 Lemur cake (composition: semolina, white cheese, butter, eggs, sweetened condensed milk, fructose, glucose, salt, and vitamins; Hladik, 1978). In addition, the animals occasionally received spiced bread, insects, vitamin-enriched milk, meat, dates, and cereals.

Food consumption in each cubicle was calculated daily according to the difference between the weight of food distributed and that of food remaining (the latter weight was obtained the next morning and multiplied by a correction coefficient corresponding to dessication; for details, see Chapter 8). From these data, the intake in terms of kilocalories was calculated from standard tables.

Observations and Records

Each cage was furnished with branches and nest-boxes, rest-platforms and feeding-platforms equipped with electrical contacts enabling the animal's activity to be recorded. Cumulative and hourly counters recorded the number of events and allowed calculation of circadian and annual activity (see Chapter 7 for technical details).

The animals were weighed monthly, genital organs examined, size of the testicles measured, and rectal temperatures were recorded with a mercury thermometer. Tail volumes of *Microcebus* and *Cheirogaleus* were estimated by measuring the volume of water displaced by immersion. During periods of sexual activity, females were examined more often, although frequent manipulations were avoided as far as possible, to ensure successful reproduction which was one of the objectives of these studies.

RESULTS

As mentioned above, photoperiodic cycles in captivity were inverse to that of Madagascar. Thus, in order to make comparisons with field studies, it is necessary to take into consideration the fact that the months of December and January (period of rest and torpor for some of the animals) correspond to June–July in Madagascar (austral winter).

Body Weight

Microcebus murinus

The *Microcebus* observed during this study showed a distinct fattening during each winter period, in 1973–1974 and 1974–1975 (Fig. 1a). Since these animals changed hemisphere in December 1972, the first months of 1973 were marked by adjustment to the new annual photoperiodic variation, the phase of which is shifted 6 months relative to the preceding one. This period of synchronization appears in Fig 1a as irregular variations in individual body weights.

During this period, the beginning of fattening which normally would have occurred for austral winter (May–June), may be clearly distinguished. However, the animals underwent a weight loss in April, which is due to a lengthening of the photoperiod. In July, synchronization with the new photoperiod seems to have been achieved. The low body weights recorded during July and August may result from either the beginning of the active period in the southern hemisphere or the end of this period in the northern hemisphere. In October all the animals showed a weight increase, in this case as a result of the decreasing photoperiod.

During the period of fattening, the volume of the tail of *Microcebus* increased from 5 to 20 cm^3. This increase of volume, however, began earlier and less abruptly than that of body weight, and also persisted during a longer period of time.

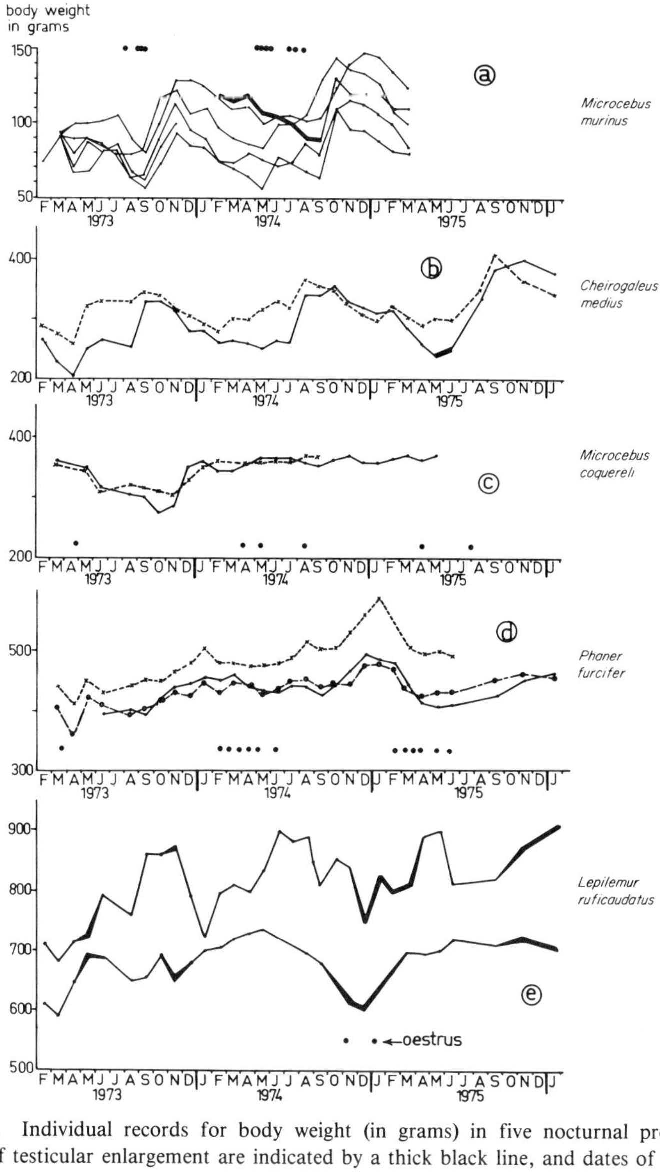

Fig. 1. Individual records for body weight (in grams) in five nocturnal prosimians. Periods of testicular enlargement are indicated by a thick black line, and dates of estrus by dots.

Cheirogaleus medius

The change of hemisphere resulted in almost complete fusion of the period of fattening (which had already begun in April, corresponding to the effect of the Malagasy photoperiodic variation) with that which normally occurs in August due to the photoperiodic variation of the northern hemisphere. This fusion (Fig. 1b) was actually facilitated by the fact that autumnal weight-gain occurs approximatively 2 months earlier than that of *Microcebus,* and may be seen by comparing the curves of body weight for the two species during the following two seasons (Fig. 1a and b).

Microcebus coquereli

The body weight of this species did not present any marked seasonal variation under our experimental conditions (Fig. 1c).

Phaner furcifer

No very distinct seasonal weight variation was observed in this species. However, slight tendency of weight increase was observed between November 1974 to January 1975, followed by a decrease from February to April (Fig. 1d). This variation occurred at the same time as those of *Microcebus* and *Cheirogaleus* and may indicate a possible physiological cycle, but of lesser importance.

Lepilemur ruficaudatus

The weight curves of the two males observed showed little variation in proportion to the total body weight (Fig. 1e). A slight decrease in body weight occurred during the months of sexual activity (November to January).

This species is distinguished from the others by the fact that the period of mating in Madagascar occurs toward the end of the period of decreasing daylength as can be seen by the period of estrus and testicular variation.

A more representative comparison between the different species can be obtained by examining the percentage monthly variation of body weight in relation to the annual mean (calculated for the 2-year period, September 1973–September 1975, except for *Microcebus murinus*) as shown in Table I. Only *Microcebus murinus* and *Cheirogaleus medius* undergo an appreciable variation: up to 20% weight loss or gain depending on the season. In *Microcebus coquereli* and *Phaner furcifer,* the variation did not usually exceed $\pm 5\%$. In *Lepilemur ruficaudatus* variations were slightly more important and suggest an annual cycle which is inverse to that of *M. murinus* and *C. medius.*

TABLE I

Mean Monthly Values for Body Weight (gm) and as a Percentage Difference from the Annual Mean in Five Nocturnal Prosimian Species

	Microcebus murinus $N = 5^a$		*Cheirogaleus medius* $N = 2^b$		*Microcebus coquereli* $N = 1^a$		*Phaner furcifer* $N = 3^b$		*Lepilemur ruficaudatus* $N = 2^b$	
Month	Weight	Percent	Weight	Percent	Weight	Percent	Weight	Percent	Weight	Percent
Jan.	117.4	+14	295	−4	359	0	493	+7	722	−5
Feb.	106.2	+3	293	−4	345	−4	473	+3	742	−3
Mar.	99.6	−3	289	−6	347	−3	459	0	760	−1
Apr.	90.8	−12	279	−10	356	0	448	−3	781	+2
May	84.8	−17	276	−11	367	+2	445	−3	794	+4
June	91.6	−11	285	−8	365	+2	445	−3	791	+3
July	90.8	−12	297	−3	364	+2	459	0	800	+5
Aug.	90.6	−12	347	+12	361	+1	469	+2	794	+4
Sept.	93.4	−9	374	+21	356	0	449	−2	758	−1
Oct.	124.8	+21	344	+11	366	+2	445	−3	777	+2
Nov.	122.0	+18	322	+4	368	+3	462	0	745	−2
Dec.	122.4	+18	305	−1	349	−2	480	+4	707	−7
Annual mean	103		309		358		460		764	

[a] During 1 year.
[b] During 2 years.

Temperatures

The recorded rectal temperatures in *M. murinus* varied between 35° and 40°C during the summer, and between 28° and 38°C during the winter. In *C. medius,* temperatures varied between 33° and 38°5C in summer and between 31° and 37°C during the winter. In all other species, rectal temperatures were maintained between 35° and 39°C throughout the year.

More recent observations concerning *M. murinus* and *C. medius* have shown that internal temperature may decrease to as low as 15°C during the winter under appropriate conditions (J. J. Petter, personal communication).

Activity

Total activity was calculated as the sum of all locomotor activity events as shown by the recording system.

The variation of the average monthly records of activity per hour was studied in relation to the yearly cycle. In order to be able to compare the results obtained from the different cubicles, regardless of the number of electrical contacts or animals present, the monthly percentage of the annual total was calculated (Table II).

In *Cheirogaleus medius,* the months of maximum activity were May, June, and July. Activity abruptly increased between April and May. Between July and August, activity suddenly decreased which corresponded to the period of rapid weight gain preceding winter (Fig. 1b).

In *Microcebus murinus,* the decrease of activity is more progressive and falls below the annual average in October. Activity is greatly reduced between November and February and increases during the summer months.

In *Phaner furcifer, Microcebus coquereli,* and *Lepilemur ruficaudatus,* the activity level does not show significant variations during the year, although there is a slight decrease during the coolest months. In *M. coquereli,* a distinct increase occurs during the period of estrus.

Food Consumption

Food consumption was calculated weekly in kilocalories, for each species between April 1973 and April 1975. The results are presented as weekly averages for each month in kilocalories per 100 gm of body weight, to allow comparison between different species (Table III).

The average monthly intake varies greatly throughout the year in *Microcebus murinus.* The minimum intake was during November to March while the maximum was in September. In *Cheirogaleus,* the maximum was in August and the minimum was attained in October.

TABLE II

Annual Variation of Activity in Five Nocturnal Prosimian Species Presented as the Monthly Total Number of Events Recorded, Mean Activity per Hour, and the Monthly Percentage of Annual Activity

	Microcebus murinus			*Cheirogaleus medius*			*Microcebus coquereli*			*Phaner furcifer*			*Lepilemur ruficaudatus*		
	Total per month	Mean per hour	Percentage per year	Total per month	Mean per hour	Percentage per year	Total per month	Mean per hour	Percentage per year	Total per month	Mean per hour	Percentage per year	Total per month	Mean per hour	Percentage per year
Jan.	475	1.1	1	876	2.5	1	12331	30.6	2	40002	9,9	4	3719	9.2	3
Feb.	1019	2.1	2	1097	3.2	2	23093	68.6	4	3869	11.5	5	5958	17.6	6
Mar.	701	2.0	2	2209	6.5	4	40293	114.7	7	5334	15.0	7	9579	27.2	10
Apr.	578	2.4	2	3038	9.4	6	59901	185.7	12	5001	15.4	7	5558	17.2	6
May	3008	9.5	9	11636	36.9	23	66823	211.9	13	6735	21.3	10	7156	22.6	8
June	5438	18.4	18	8292	44.3	28	47829	256	16	6069	20.5	16	10185	34.5	12
July	6172	20.6	21	9001	30.0	19	36077	120	7	7609	25.3	11	12492	41.7	15
Aug.	5183	16.7	17	2865	9.2	6	53722	173	11	5004	16.1	7	7656	24.7	10
Sept.	4399	13.9	14	1787	5.6	3	37989	120	8	7395	23.4	11	8414	26.7	10
Oct.	2834	8.3	8	1110	3.3	2	41718	122.7	8	8246	24.2	11	8286	24.3	9
Nov.	1045	3.0	3	1102	3.2	2	45448	130	8	6998	20.0	9	7061	20.0	7
Dec.	650	1.7	2	948	2.5	1	23749	62.5	4	6074	16.0	7	3676	9.6	3

TABLE III
Mean Weekly Food Consumption, in Kilocalories per 100 gm of Body Weight Calculated for Each Month of the Year[a]

	Microcebus murinus				*Cheirogaleus medius*				*Microcebus coquereli*				*Phaner furcifer*				*Lepilemur ruficaudatus*			
Month	1973	1974	1975	Monthly mean	1973	1974	1975	Monthly mean	1973	1974	1975	Monthly mean	1973	1974	1975	Monthly mean	1973	1974	1975	Monthly mean
Jan.		50	52	51		47	54	50		121	115	118		78	98	88		66	73	69
Feb.		(60)	68	64		(60)	48	54		(134)	127	130		(84)	91	87		(77)	74	75
Mar.		89	79	84		100	40	70		128	141	134		88	86	87		67	71	69
Apr.	122	118	109	116	104	84	55	81	104	158	151	138	92	98	99	96	73	82	78	78
May	117	141		129	102	92	106	100	143	125		134	97	94		95	74	70		72
June	(67) ±	126		126	(52) ±		134	134	(97) ±	114		114	(84) ±	93		93	(47) ±	67		67
July	85 ±				104		163	135	127		186	156	102			102	58			53
Aug.	141		220	180	161		201	181	150		184	167	111		128	119	92		92	92
Sept.	(228)	235	246	239	(62)	54	105	74	(194)	155		174	(95)	91	132	106	(76)	73	91	80
Oct.	177	179		178	54	57		55	139			139	91	98		94	59	95		83
Nov.	65	56		60	49	48		48	151	159		150	106	118		112	52	57		54
Dec.	32	51		41	54	45		49	104	142		123	102	113		107	44	60		55
Annual mean (kcal/week/100 gm body weight)				115				86				140				99				71

[a] Values in parentheses were calculated from only 1 or 2 weeks of the month. Values with ± were considered abnormal, due to extraneous factors and were not included in mean calculation.

In *Lepilemur ruficaudatus, Phaner furcifer,* and *Microcebus coquereli,* food consumption varies little. However, there is a slight increase in food intake between July and October, and a decrease during the coolest months.

There are important differences in the annual mean consumption in kilocalories per week per 100 gm (Table III). The highest value is that of *M. coquereli* (140 kcal), while the lowest is that of *L. ruficaudatus* (71 kcal). This difference corresponds to both the general activity level of the animals, (the former being the most active, the latter being the least active) and the difference in body weight. In general, the larger species show a lower rate of food consumption corresponding to a relatively lower metabolism. However, the two species which show periods of lethargy (*M. murinus* and *C. medius*) have annual values lower than expected according to body weight.

Sexual Activity

The occurrence of estrus and variations in the size of testicles were observed in all species (Fig. 1). Only *Cheirogaleus medius* showed an incomplete sexual cycle.

Microcebus murinus

Estrus occurred between May and August in the four females studied during 1974. Normal testicular development in the male was observed between April and September. This sexual cycle corresponds to previous observations of *Microcebus* in captivity, and has already been described in detail (Petter-Rousseaux, 1970). A birth occurred in July 1976.

Cheirogaleus medius

In *C. medius,* only slight testicular development was observed in the male (May–June, Fig. 1b). During the 3 years of observations, no sexual cycle was noted in the female.

Microcebus coquereli

During the period of sexual activity, the external genital organ of *M. coquereli* is similar in morphology to that of *M. murinus* (lengthened clitoris, vulva completely closed). During estrus the same modifications as in *M. murinus* may be observed: swelling and reddening at the base of clitoris, followed by opening of the vulva for several days.

The first estrus appears in April; a second estrus may follow about a month later if gestation does not occur. The period of gestation lasts for 3 months. If the newborn does not survive, another estrus may occur the following month. Births took place on the following dates:

1973: August 1
1974: July 1, and November 1
1975: July 14, and November 20

In males, distinct variations in the length of the testicles were observed, and the period of regression lasted only for 1 or 2 months.

Phaner furcifer

In one of the females, estrus occurred only once per year. The other female presented a series of four or five estrus per year (Fig. 1d). Spermatozoid traces were twice found in the vaginal smears.

The external aspect of the genital organs was the same as that of the other Cheirogaleinae studied. Estrus began toward January 15. The vulva remains open for only 2 or 3 days and the preparatory period of swelling and reddening was equally very limited. Estrus occurred at fairly regular intervals of about 15 days.

In the male, testicular variation is slight and difficult to evaluate reliably. However, the testicles appear to increase in size between January and March.

Lepilemur ruficaudatus

The two *Lepilemur* females arrived in March 1974, and thus it was only possible in September to form two couples with the two males which were in captivity since January 1973.

After their arrival the females were maintained in conditions of long daylength in order to inhibit their sexual activity which would have developed in May under the Malagasy photoperiod. The females were introduced into the males' cages during the period of decreasing daylength, the normal period for mating. One of the females showed two periods of estrus in November and in January.

Estrus is characterized by the swelling and the pinkening of the edges of the vulva, which opens widely. As in Cheirogaleinae, the vagina is closed during the period of sexual inactivity. The occurrence of estrus in winter, (Fig. 1e), is corroborated by previous observations of a female in natural light conditions of the northern hemisphere, in which estrus occurred on December 19 and February 10.

In the male, there is a distinct variation in the size of the testicles. However, during the 3 years of the experiment, the volume never attained that which is observed in nature. The testicles develop between November and March, thus during the season opposite to that of Cheirogaleinae. During the first year of captivity, a slight testicular development was noticed in

May, corresponding to that which would have normally occurred in Madagascar. Mating was observed on November 9, 1974.

A birth took place the year following the experiment, on June 2, 1976.

DISCUSSION

The five species studied in captivity were raised in an environment providing less overall annual variations than those which normally occur in nature. Although lighting could be accurately controlled, modifications of temperature and humidity were incompletely reproduced.

In addition, food availability and composition were roughly constant throughout the year, a significant contrast from field conditions: this fact must thus be taken into consideration for interpretation of the results. For example, the regular food supply and possibly the high nocturnal temperatures in winter, may have been responsible for the high weight gains observed in *Microcebus murinus* and *Cheirogaleus medius* compared to field records.

The social life of the animals was certainly also disturbed by the limited space. In their natural habitat, the home range varies from 1000 to 4500 m^2 depending on the species (see Chapters 2 and 4). However, neither the small size of the cages nor the proximity of other individuals resulted in abnormal behaviors of the animals during captivity.

The mean monthly results concerning weight variations, food intake, and general activity, for the five species are summarized in Fig. 2.

Two species, *Microcebus murinus* and *Cheirogaleus medius,* may be distinguished by the presence of distinct physiological cycles, concerning food intake, body weight, temperature regulation, and activity. In addition, only these two species show definite periods of "lethargy" and/or "hibernation." In *M. murinus* the period of lethargy, observed both in the laboratory and in the field (during austral winter), is preceded by increased food intake and the accumulation of fat reserves in the body and the tail (30% increase in body weight in 1 month), and a decrease in activity. Subsequently, lethargy is characterized by extremely reduced activity, decreased food intake, and the possibility of reduced body temperature during the day and eventually for several consecutive days and nights. Although this last feature was not observed during the present study, *M. murinus* has recently been shown to remain totally inactive for periods of 2–3 days during which body temperature decreased (minimum body temperature recorded: 15°C).

Cheirogaleus medius also shows these variations, but the physiological changes preceding lethargy occur 2 months earlier than in *M. murinus.* The

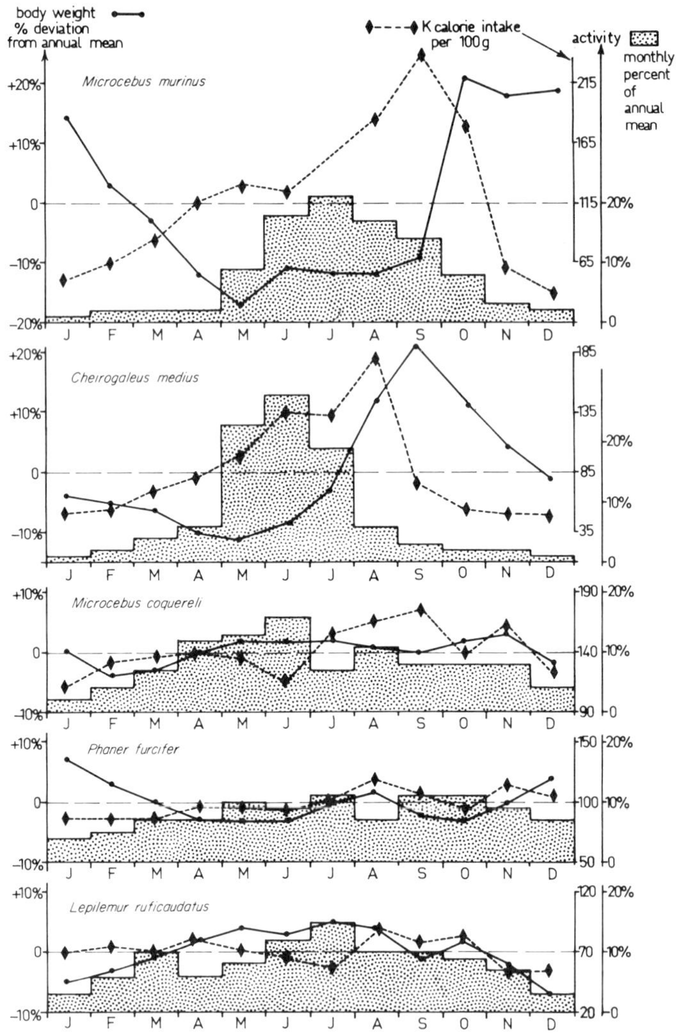

Fig. 2. Mean monthly variations in body weight (percentage deviation from the annual mean; solid line), food consumption (kcal/week/100 gm of body weight, dashed line), and of general activity (percentage of total annual activity, stippled area) in five prosimian species.

decrease in activity was more abrupt, but the increases in food intake and body weight were more progressive. However, in the case of *Cheirogaleus medius* a phase of hibernation occurs which has been observed in the Marosalaza forest (Chapter 2). This phase was not obtained during the present experiment, probably due to insufficiently low temperature during winter. In fact, since the time of this study normal hibernation (and also births) have been observed in *Cheirogaleus medius* subjected to winter temperatures as low as 12°C. The animals were totally inactive for 6 months, and body temperatures were close to that of the external environment (J. J. Petter, personal communication).

Previous studies concerning *Microcebus murinus* and *Cheirogaleus medius* (Bourlière *et al.*, 1956; Russell, 1975; Andriatsiferana and Rahandraha, 1974; Chevillard, 1976) have also shown that body temperatures can decrease to very low values in appropriate conditions.

Thus, these two Cheirogaleinae resemble, in certain respects, hibernating mammals in which sudden and marked weight gains are well documented, e.g., *Citellus* (Pengelley and Asmundsen, 1969); *Spermophilus lateralis* (Heller and Poulson, 1970); *Marmotta monax* (Davis, 1967); *Cricetus cricetus* (Canguilhem *et al.*, 1973) and for which it has been shown that physiological rhythms also depend on external factors, which have a synchronizing role, such as the photoperiod or temperature. As in the case of these species, the physiology of *M. murinus* and *C. medius* enables them to efficiently face an unfavorable season characterized by low temperature and food scarcity.

The three other species, *Microcebus coquereli, Phaner furcifer,* and *Lepilemur ruficaudatus,* do not present any period of lethargy. The slight annual variations observed, which may be more pronounced in the natural environment due to seasonal food scarcity, did not show the characteristic peaks seen in the two previous species, and thus suggests a different type of physiological response to seasonal changes.

The species sexual cycles were synchronized by the photoperiod. All species reversed their period of sexual activity once subjected to the photoperiodic variations of the northern hemisphere. The phase of adaptation lasted approximately 6 months (January to July 1973). This is best illustrated by *Lepilemur ruficaudatus* in which testicular development occurred in May–June, and again in November–December 1973 (Fig. 1e). The behavior of the different species during this period, as well as the duration of the period varies with the individual, and depends on the season during which the hemisphere change occurs (Petter-Rousseaux, 1974).

The periods of mating (corresponding to periods of estrus and testicular enlargement, see Fig. 1) and the duration of the period of gestation result in that the young of all species begin to feed when food resources are most abundant in natural conditions (austral summer). For example, in *Lepilemur ruficaudatus* which has the longest gestation period (4–5 months) the mating season occurs earlier than that of the other species.

The persistence of the various cycles in simulated light conditions demonstrates that they are not simple reactions to short-term changes in food and climatic conditions. The photoperiod appears to be the synchronizing factor whereas other external parameters (particularly temperature) condition the full expression of physiological reactions externally manifested by activity and torpor, and internally by sexual maturation at a given time.

These different genetically fixed potential reactions play a major role in each species adaptive strategy to the contrasting seasonal conditions of the West Coast of Madagascar.

REFERENCES

Andriatsiferana, R., and Rahandraha, T. (1974). Effets du séjour au froid chez le Microcèbe (*Microcebus murinus*). *C. R. Hebd. Seances Acad. Sci., Ser. D* **278,** 3099–3102.

Bourlière, F., Petter, J. J., and Petter-Rousseaux, A. (1956). Variabilité de la température centrale chez les Lémuriens. *Mem. Inst. Sci. Madagascar, Ser. A* **10,** 303–304.

Canguilhem, B., Schieber, J. P., and Koch, A. (1973) Rythme circannuel pondéral du Hamster d'Europe (*Cricetus cricetus*). *Arch. Sci. Physiol.* **27,** 67–90.

Chevillard, M. C. (1976). Capacités thermorégulatrices d'un Lémurien malgache *Microcebus murinus* (Miller 1777). Thesis (3rd cycle), University of Paris VII.

Davis, D. E. (1967). The annual rhythms of fat deposition in Woodchucks (*Marmotta monax*). *Physiol. Zool.* **40,** No. 4, 391–402.

Heller, H. C., Poulson, T. L. (1970). Circadian rhythms. II. Endogenous and exogenous factors controlling reproduction and hibernation in Chipmunks (*Eutamias*) and ground squirrels (*Spermophilus*). *Comp. Biochem. Physiol.* **33,** 327–383.

Hladik, C. M. (1978). Diet and ecology of prosimians. *In* "The Study of Prosimian Behavior" (G. A. Doyle and R. D. Martin, eds.), pp. 307–357. Academic Press, New York.

Pengelley, E. T., and Asmundsen, S. M. (1969). Free-running periods of endogenous circadian rhythms in the Golden mantled Ground Squirrel (*Citellus lateralis*). *Comp. Biochem. Physiol.* **30,** 177–183.

Perret, M. (1974). Variations of endocrine glands in the Lesser Mouse-Lemur *Microcebus murinus*. *In* "Prosimian Biology" (R. D. Martin, G. A. Doyle, and A. C. Walker, eds.), pp. 375–387. Duckworth, London.

Perret, M. (1975). Activité thyroïdienne de *M. murinus* (Miller 1977). Influence de la captivité. *Mammalia* **39,** 119–132.

Petter, J. J. (1978). Ecological and physiological adaptation of five sympatric Lemurs to seasonal variations in food production. *In* "Recent Advances in Primatology" (D. J. Chivers and J. Herbert, eds.), Vol. 1, pp. 211–223. Academic Press, New York.

Petter, J. J., Albignac, R., and Rumpler, Y. (1977). Mammifères Lémuriens (Primates Prosimiens). *Faune Madagascar* **44.**

Petter-Rousseaux, A. (1970). Observations sur l'influence de la photopériode sur l'activité sexuelle de *Microcebus murinus* en captivité. *Ann. Biol. Anim., Biochim., Biophys.* **10** (2), 203–220.

Petter-Rousseaux, A. (1974). Photoperiod, sexual activity and body weight variations of *Microcebus murinus* (Miller 1777). *In* "Prosimian Biology" (R. D. Martin, G. A. Doyle, and A. C. Walker, eds.), pp. 665–673. Duckworth, London.

Russell, R. J. (1975). Body temperatures and behavior of captive Cheirogaleids. *In* "Lemur Biology" (I. Tattersall and R. W. Sussman, eds.), pp. 193–206. Plenum, New York.

7 Annual Variations in the Circadian Activity Rhythms of Five Sympatric Species of Nocturnal Prosimians in Captivity

E. PAGES and A. PETTER-ROUSSEAUX

INTRODUCTION

The five species of nocturnal prosimians compared in the present study (*Microcebus murinus, Microcebus coquereli, Cheirogaleus medius, Phaner furcifer,* and *Lepilemur ruficaudatus*) exist sympatrically in the deciduous forest of Marosalaza, on the West Coast of Madagascar.

The climate of this region is marked by a hot and humid summer (average temperature, 28°C), during which the vegetation and insect populations flourish, and a cool dry winter (average temperature, 20°C) during which animal and vegetal food resources are greatly reduced (see Chapter 1).

In order to understand both the mechanisms of each species adaptation to the conditions of seasonal variation and the extent of interspecific competition, complementary ecophysiological investigations were undertaken both in the field and in the laboratory. Part of this study concerning reproduction, activity rhythms, and body weight is reported in Chapter 6.

Initial field studies have shown that activity rhythms depend on external factors, such as light intensity (Pariente, 1974) but that variations throughout the night during different seasons may characterize certain species (Chapter 4).

Accordingly, the present study was directed toward answering the following questions:

1. Are the five species strictly nocturnal in activity throughout the entire year?
2. Are there variations of activity during the night, and do these patterns vary within and between species during the yearly cycle?

3. Are the species cycles in captivity similar to those observed in the field?

4. What factors determine the onset and cessation of activity?

METHODS

Housing and Climate Control

During a period of 2 years, observations and recordings were made on the five species housed in the same room. Each cage, which measured about 3 × 2 × 2 m, was constructed with glass walls and wire mesh roofs permitting visual, auditory, and olfactory contacts.

The various species were distributed in thc cages according to social structures and inter-individual tolerance, in the following manner:

1. *Microcebus murinus:* one male and four females
2. *Microcebus coquereli:* one male–female pair with a young, and one isolated male in an adjoining cage
3. *Cheirogaleus medius:* one male–female pair
4. *Phaner furcifer:* one male and two females
5. *Lepilemur ruficaudatus:* two males only in 1973–74; two male–female pairs in 1974–75.

The climatological changes recorded in the Marosalaza forest (Chapter 1) were reproduced in the animal house as much as was practically possible, with a 6-month difference. Variations in the photoperiod were controlled by electric timers including the half hour periods of "dawn" and "dusk" (see Chapter 6 for details).

During the artificial night period, 25-W red lamps provided sufficient light to make observations possible without disturbing the animals.

Temperature was regulated by thermal heaters (during the day) and by thermostat-controlled ventilators (during the night). This system allowed control of daily temperature cycles, and produced a mean of 28°C during the day, and 25°C at night in summer, while in winter, day and night temperatures were 27° and 21°C, respectively.

The relative humidity remained fairly low (as compared to field conditions), with a mean of 42% in summer and 24% in winter.

Recording of Activity

The animals' activities were automatically recorded by electrical microswitches located at specific places within each cage (Fig. 1). When the animal's movement was detected by closure of one of the microswitches,

TABLE II

Time of Cessation of Activity (Mean and Standard Deviation in Hours/Minutes) in Five Nocturnal Prosimians

Month	Time of lighting on	*Microcebus murinus*		*Cheirogaleus medius*		*Microcebus coquereli*		*Phaner furcifer*		*Lepilemur ruficaudatus*	
Jan.	22:25–22:50					Incomplete data		21:40	0.37	21:32	0.32
Feb.	22:30–22:50	20:40	0.44			Incomplete data		21:40	0.17	21:40	0.41
Mar.	22:00–22:30	20:15	1.03	20:15	1.06	21:40	0.28	21:45	0.17	21:40	0.24
		21:40	0.26	21:40	0.17						
Apr.	21:30–22:00	21:40	0.25	21:10	0.26	21:15	0.18	21:20	0.18	21:25	0.06
May	21:10–21:40	21:30	0.12	20:55	0.18	21:00	0.13	21:20	0.20	21:20	0.10
June	21:10–21:30	21:30	0.24	21:00	0.12	21:00	0.21	20:55	0.20	21:45	0.51
July	21:10–21:40	21:35	0.07	21:15	0.16	21:15	0.20	21:15	0.23	21:45	0.18
Aug.	21:30–22:00	21:45	0.13	21:15	0.15	21:35	0.15	21:10	0.25	21:40	0.10
Sep.	21:50–22:20	21:50	0.33	21:00	0.44	21:40	0.20	21:45	0.16	21:40	0.45
Oct.	22:00–22:30	21:20	0.54	21:10	0.54	21:25	0.25	21:40	0.15	21:50	0.57
Nov.	22:10–22:40	20:05	1.58	20:55	0.42	21:10	0.49	21:30	0.23	22:00	0.24
Dec.	22:20–22:45					21:30	0.41	21:45	0.50	22:05	0.43

From October to March, all species returned to the nest before morning twilight. However, *Microcebus murinus* and *Cheirogaleus medius* returned well before morning dawn and at irregular hours. Toward March 15, this behavior suddenly changed to resemble that of the other three species.

Duration of Nocturnal Activity (Figs. 2 and 3; Table III)

As a result of the above activity rhythms, the duration of activity varied throughout the year according to the total time of darkness. This appears in Fig. 3 in terms of percentage of the nighttime during which animals were active. In summer, animals were active at night and a part of the twilight as well. Conversely, in winter, the animals were not active for the entire night. In all species, the percentage of the nighttime utilized decreases in October–November, as a result of the nights becoming longer while the active period of the animals was approximately constant. In November, after the night length is considerably increased, *Microcebus murinus* and *Cheirogaleus medius* entered into a period of lethargy whereas the three other species extended their activity.

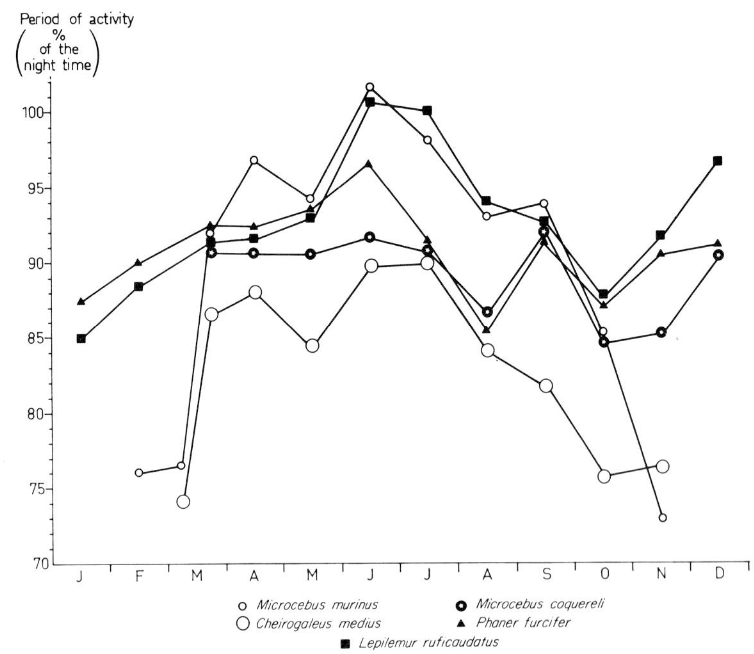

Fig. 3. Length of activity for five nocturnal prosimians, as a percentage of the total nighttime period throughout the year.

TABLE III

Mean Length of Period of Activity (with Standard Deviation in Hours/Minutes) and Percentage of the Nighttime Period in Five Nocturnal Prosimian Species

Month	Nighttime period (twilight added) (hours)	*Microcebus murinus*	*Cheirogaleus medius*	*Microcebus coquereli*	*Phaner furcifer*	*Lepilemur ruficaudatus*
Jan.	13:00			Incomplete data	11:40 ± 0.33 87.7%	11:05 ± 0.39 85%
Feb.	12:50	9:50 ± 0.39 76%		Incomplete data	11:25 ± 0.24 90%	11:05 ± 0.42 88.4%
Mar.	12:20	9:35 ± 1.23 76.6%	9:05 ± 3.03 74.2%	11:15 ± 0.27 91.4%	11:25 ± 0.19 92.2%	11:20 ± 0.25 91.8%
		11:20 ± 0.32 91.8%	10:55 ± 0.22 86.5%			
Apr.	11:45	11:10 ± 0.31 96.9%	10:10 ± 0.30 88.2%	10:40 ± 0.25 90.8%	10:55 ± 0.28 92.1%	10:50 ± 0.16 91.7%
May	11:10	10:45 ± 0.18 94.1%	9:40 ± 0.21 84.7%	10:05 ± 0.16 90.5%	10:40 ± 0.25 93.7%	10:35 ± 0.20 93.2%
June	10:40	10:50 ± 0.30 101%	9:35 ± 0.28 89.9%	9:55 ± 0.21 91.8%	10:05 ± 0.24 96.6%	10:50 ± 0.53 101%
July	10:40	10:20 ± 0.10 98.1%	9:40 ± 0.28 90.4%	9:40 ± 0.34 90.4%	9:50 ± 0.21 91.3%	10:40 ± 0.35 100%
Aug.	11:00	10:30 ± 0.14 93.6%	9:25 ± 0.18 84.1%	9:55 ± 0.21 86.8%	9:40 ± 0.28 85.4%	10:30 ± 0.20 93.6%
Sep.	11:30	10:55 ± 0.24 93.4%	9:25 ± 0.36 81.9%	10:35 ± 0.24 91.6%	10:35 ± 0.18 91.6%	10:50 ± 0.29 92.9%
Oct.	12:00	10:20 ± 0.53 85%	9:10 ± 1.17 75.8%	10:15 ± 0.27 84.6%	10:45 ± 0.15 87.1%	10:55 ± 0.18 87.9%
Nov.	12:40	9:05 ± 2.48 73%	9:45 ± 0.39 76.2%	10:55 ± 0.51 85.1%	11:25 ± 0.22 90.7%	11:40 ± 0.34 91.9%
Dec.	12:55			11:35 ± 0.36 90.4%	11:45 ± 0.37 91.2%	12:15 ± 0.46 96.8%

The longest periods of activity of *Lepilemur ruficaudatus* occurred in December–January, during the mating season. This explains the high percentage of nocturnal activity as compared to other species (Fig. 3). After March, the shorter nights appeared to impose a limit to the length of activity.

Distribution of Nocturnal Activity (Fig. 4)

Locomotor activity and feeding activity were computed separately since they did not follow the same patterns of nightly variations in different species.

In *Microcebus murinus,* there was little variation in the hourly distribution of activity during different periods of the year (Fig. 4a). Locomotor activity remained equally important throughout the night. In contrast, feeding activity presented a maximum at the beginning of the night. This pattern was still apparent in July, when feeding activity considerably increased.

In *Cheirogaleus medius,* locomotor activity as well as feeding tended to be increased at the end of the night (Fig. 4b).

In *Microcebus coquereli,* the most noticeable characteristic of the activity graphs is the asymmetrical distribution during the period April to November (Fig. 4c). During this time, locomotor activity was concentrated during the first 6 hours of the night. The reduction of locomotor activity during the second half of the night did not actually correspond to a period of inactivity. During this time, the animal engaged in social activities as has been observed in captivity as well as in the field (Chapter 4). In January, when there was a minimum of overall activity (Chapter 6), this asymmetry was no longer apparent. As for *Microcebus murinus,* there was a maximum of feeding activity at the beginning of the night.

In *Phaner furcifer,* the most important peak of locomotor activity also occurred during the first half of the night (Fig. 4d). This activity profile persisted throughout the year, but was slightly reduced in intensity from November to February. Feeding activity often followed an opposite pattern, but with only a slight difference between the beginning and end of the night.

In *Lepilemur ruficaudatus,* locomotor activity varied little throughout the night (Fig. 4e). Feeding activity was most important in the first hours of the night, and progressively decreased afterwards.

For any species, the overall activity may vary greatly from day to day and does not seem to follow any particular periodicity. Occasional peaks of activity often involved all the animals and might be explained by the easy intercommunication between cages. Nevertheless, the monthly means

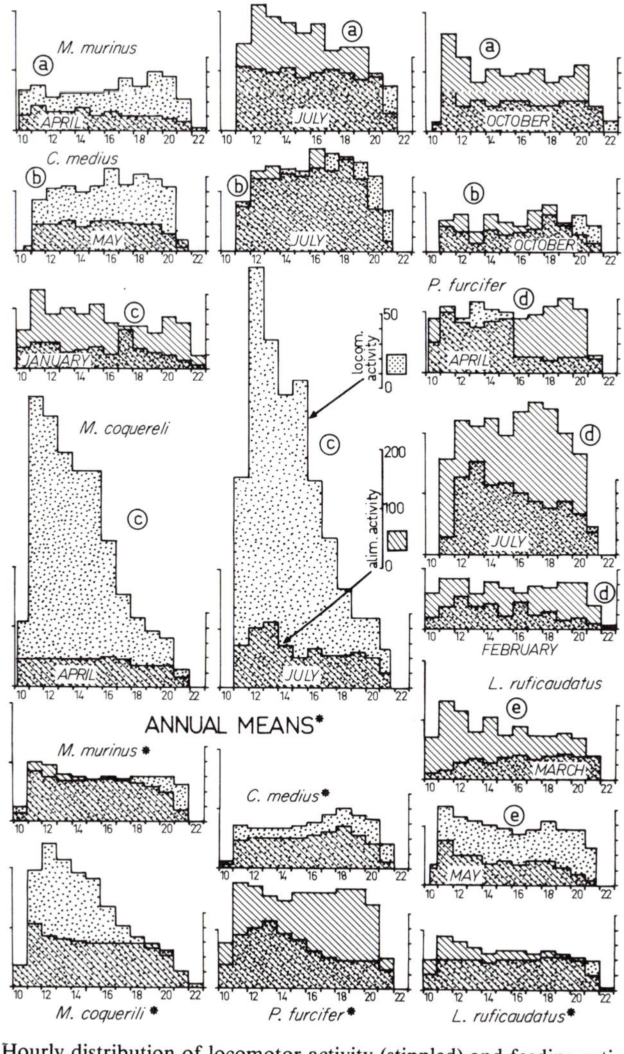

Fig. 4. Hourly distribution of locomotor activity (stippled) and feeding activity (slashed) in five nocturnal prosimians, as shown by the number of individual microswitch recordings (due to particular distribution of microswitches, the units of feeding activity cannot be compared to those of locomotor activity). The annual mean for the different species is shown (stars) as well as the most characteristic patterns of activity (monthly average).

presented in Fig. 4 show significant interspecific differences which are related to behavioral and/or physiological characters.

DISCUSSION

A strictly nocturnal rhythm of activity characterizes the five prosimian species included in the present study. The African lorisids have also been found to be active uniquely at night (Kavanau and Peters, 1976). In this respect, nocturnal prosimians differ from many rodents and insectivores which are mainly nocturnal, but can facultatively adopt a diurnal activity according to the season (O'Farrel, 1974).

The beginning of activity at dusk was very regular in the five prosimians, and occurred shortly after sunset, whereas cessation of activity (which was more variable) occurred well before dawn. The synchronization of the animals' activity to variations in the photoperiod thus appears to be triggered by the evening twilight. Indeed, if differences in light intensity have a synchronizing role (Bunning, 1960), then in the case of prosimians, this effect is likely to occur at evening dusk.

Important increases in activity occurred at the time of estrus, particularly during the first half of the night, whereas no distinctive peaks were observed in the present experimental conditions during the rest of the year. Nevertheless, important peaks of activity have been observed in different species, in nature as well as in artificial conditions. Among rodents such peaks vary, depending on species and season (O'Farrel, 1974), and may persist in the absence of day–night light cycle synchronization (Chow and Brown, 1975). Although these rhythms thus appear to be "internally programmed," they may be influenced by changes in social status (e.g., *Peromycus,* Farr, 1975; *Rattus,* Calhoun, 1975). In primates, Trach (1969) presented evidence for an "internal rhythm" in *Papio papio,* since activity remained bimodal in constant illumination.

Patterns of activity shown for the five prosimians in this study, have also been observed in the field. In *Microcebus murinus,* Martin (1973) observed that most activity occurred in the first and last third of the night. In *Lepilemur* sp. periods of immobility lasting more than 2 hours were noted by Charles-Dominique and Hladik (1971). In contrast, *Phaner furcifer* is particularly active at the beginning of the night (Petter *et al.,* 1971; Chapter 3). *Microcebus coquereli* engages in different types of activities during the first and second half of the night, and includes rest periods which may last up to 1 hour (Chapter 4).

Among the African Lorisidae, *Galago demidovii* presents two distinct peaks of activity, whereas the activity of *Galago elegantulus* and *Perodic-*

ticus potto is more evenly distributed throughout the night (Charles-Dominique, 1971, 1977).

In diurnal prosimians, the bimodal distribution of activity appears to be dependent on the important variation of temperature during the daytime with a rest period in the midday, for instance, in *Lemur catta* (Jolly, 1966), *Lemur macaco,* and *Lemur fulvus* (Petter *et al.,* 1977). Conversely, the regular decrease of the night temperature may impose a decrease in the activity in certain nocturnal species.

Other external factors such as the cycle of the moon, or precipitations, may influence activity (Marten, 1973). In addition, internal factors also may be related to the activity rhythm, for instance, social motivations in *M. coquereli* (Pages, in prep.).

During the present comparative study of the five Malagasy prosimians, a marked decrease in activity occurred in most cases between the first half and the second half of the night, but periods with complete absence of activity rarely occurred (except for *C. medius* and *M. murinus* during winter lethargy). This was probably due to the restricted space available in the cages since even limited movements were recorded. The same type of differential bias also exists for feeding activity records, due to the location of the feeding platform, which results in individuals feeding successively instead of simultaneously, thus increasing the duration of the feeding activity.

Nevertheless, the results provide supplementary indications concerning the occurrence of lethargy and hibernation in *M. murinus* and *C. medius* (Chapter 6). Beginning in October/November, the onset of activity no longer follows twilight (Fig. 2), resulting in a reduction of overall activity. In March the activity of both species coincides again with dusk.

The specific patterns of activity found in the different nocturnal prosimians are not due to the environmental parameters of food availability since, in the present work, these conditions have been standardized (Chapter 8). Nevertheless, many relationships with the different "strategies" used by each species to cope with the variations in food availability in the natural habitat can be shown when comparing the results obtained in captivity and from field observations (see Chapters 2, 3, and 4).

Feeding activity was more important at the beginning of the night in the two *Microcebus* species, while in *C. medius,* feeding behavior often occurred at the end of the night. Furthermore, during the active period (e.g., in July, Fig. 4) activity of *C. medius* was significantly more important during the second half of the night. This overall activity pattern is exactly the reverse of the other species (except perhaps *L. ruficaudatus*).

It is noteworthy that *M. murinus* and *C. medius* are potential competitors, especially for invertebrate food. Different times of feeding may greatly reduce competition. In addition to this temporal separation, the significant difference in terms of the height of feeding also accounts for niche separation (Chapter 2).

In *M. coquereli* as well as in *Phaner furcifer,* the patterns of distribution of activity throughout the night and their respective seasonal variations are also dependent on the particular social organization (Chapters 3 and 4). These patterns must thus be interpreted not only in terms of interspecific competition, but also in relation to intraspecific factors relating to ecology and social behavior.

REFERENCES

Bunning, E. (1960). Opening address: Biological clocks. *Cold Spring Harbor Symp. Quant. Biol.* **25,** 1–9.

Calhoun, B. (1975). Social modifications of activity rhythm in rodent. *Chronobiologia, Suppl.* **1,** 11. (abstr.).

Charles-Dominique, P. (1971). Eco-éthologie des Prosimiens du Gabon. *Biol. Gabonica* **7,** No. 2, 121–228.

Charles-Dominique, P. (1977). Ecology and Behaviour of Nocturnal Primates." Duckworth, London.

Charles-Dominique, P., and Hladik, C. M. (1971). Le Lepilemur du sud de Madagascar: Ecologie, alimentation et vie sociale. *Terre Vie* **1,** 3–66.

Chow, C. S., and Brown, F. A., Jr. (1975). Entrainment to 24 hrs of free-running Hamster circadian rhythms in darkness by a subtle field Zeitgeber. *Chronobiologia, Suppl.* **1,** 24 (abstr.).

Farr, L., and Andrews, R. V. (1975). Social rank effects on metabolic, feeding and activity profiles of deer-mice. *Chronobiologia, Suppl.* **1,** 38 (abstr.).

Jolly, A. (1966). "Lemur Behavior. A Madagascar Field Study." Univ. of Chicago Press, Chicago, Illinois.

Kavanau, J. L., and Peters, C. R. (1976). Activity of nocturnal Primates; influences of twilight Zeitgebers and weather. *Science* **191,** 83–86.

Marten, G. G. (1973). Time patterns of *Peromyscus* activity and their correlation with weather. *J. Mammal.* **54,** 169–188.

Martin, R. D. (1973). A review of the behaviour and ecology of the Lesser Mouse Lemur (*Microcebus murinus* J. F. Miller 1777). *In* "Comparative Ecology and Behaviour of Primates" (R. P. Michael and J. H. Crook, eds.), pp. 1–68. Academic Press, New York.

O'Farrel, M. J. (1974). Seasonal activity patterns of Rodents in a sagebrush community. *J. Mammal.* **55,** 809–823.

Pariente, G. F. (1974). Influence of light on the activity rhythms of two Malagasy Lemurs: *Phaner furcifer* and *Lepilemur mustelinus leucopus. In* "Prosimian Biology" (R. D. Martin, G. A. Doyle, and R. C. Walker, eds.), pp. 183–198. Duckworth, London.

Petter, J. J., Schilling, A., and Pariente, G. (1971). Observations éthologiques sur deux Lémuriens malgaches nocturnes, *Phaner furcifer* et *Microcebus coquereli. Terre Vie* **3,** 287–327.

Petter, J. J., Albignac, R., and Rumpler, Y. (1977). Mammifères Lémuriens. *Faune Madagascar* **44.**

Trach, J. S., Jr. (1969). A bigeminus pattern in social behavior. *In* "Circadian Rhythms in Nonhuman Primates" (F. H. Rohles, ed.), Bibli. Primatol. No. 9. Karger, Basel.

8 A Comparative Study of Food Intake in Five Nocturnal Prosimians in Simulated Climatic Conditions

A. PETTER-ROUSSEAUX and C. M. HLADIK

INTRODUCTION

The *Morondava Programme* has been developed to elucidate problems of adaptation concerning five nocturnal prosimian species living sympatrically in the dry forest of the West Coast of Madagascar (Petter, 1978; Chapter 2). Research at the laboratory of Brunoy was undertaken to supplement the field studies with accurate measures of certain parameters in standard conditions, in order to understand the mechanisms and processes of regulation of species adaptation. The success of the breeding colony allowed data collection on a long-term basis, concerning the five species: *Microcebus murinus, Microcebus coquereli, Cheirogaleus medius, Phaner furcifer,* and *Lepilemur ruficaudatus.* During 3 consecutive years, body weight, body temperature, locomotor activity, and sexual activity were continually observed (Chapter 6). The annual variation in the circadian rhythm of the different species has been analyzed and compared over a period of 2 years (Chapter 7).

In the present study, we attempted to define the overall dietary tendencies of each species by adopting standard methods of comparison of food choices. In natural conditions, these tendencies are expressed against a highly variable background of the forest cycles of phenology and production (Chapter 1) and are thus difficult to quantify. The present study in artificial conditions was necessary to determine whether species manifest characteristic food choices, and to what extent dietary choices vary throughout the year, independent of variations in food availability.

METHODS

The animals were maintained in large cages constructed of glass and wire mesh, described in Chapter 6, and grouped according to intraspecific tolerance: five *Microcebus murinus,* three *Microcebus coquereli,* two *Cheirogaleus medius,* three *Phaner furcifer,* and two *Lepilemur ruficaudatus* in each of the cages.

The climatic conditions in the animal house (see Chapter 6) approximated those which had been recorded in the Marosalaza forest, West Coast of Madagascar (see Chapter 1), with a 6-month difference. Accordingly, the summer months in France (June, July, August) corresponded to austral summer in Madagascar (December, January, February).

Food intake was recorded over a period of 2 years according to the following procedure: A set of standard foods including salad and other leaves, different fruits, meal worms and other insects, and a protein-rich food [a cake of semolina and animal protein described by Hladik (1978a) as the G3 Lemur cake, or in some instances a mixture of milk and cooked cereals] was provided in each of the cages, daily, before the period of artificial night. On weekends a quantity of food sufficient for 2 days was given. Water was always available in each cage.

The different components of these "sets of foods" were separately weighed, as well as a control set. This control set served as a reference for food dessication, and was placed in the animal house adjacent to the cages. Before daily feedings, all food remaining from the previous day in the different cages was carefully collected and weighed. The control set of food allowed calculation of the loss of water of the different components. From this data, the exact fresh weight ingested by the animals in the different cages could be calculated.

Samples of the different foods were separately collected and dried in an electric oven at 60°C, to allow further biochemical analysis.

During the 2-year period of experimentation the same set of food was provided (whenever possible) in each of the cages, and always slightly in excess of the actual amount ingested, in order to allow each species to manifest its particular choice. Nevertheless, there were some variations depending on the fruits available seasonally, and the insect supply was not sufficient to provide each cage at regular intervals. In addition, during certain periods of the year, the G3 Lemur cake could not be maintained in excess of consumption. Accordingly, the comparisons were mainly based on the gross categories of food constantly available, which represent fairly homogeneous groups in terms of biochemical composition.

The variations in food supply which could not be avoided were taken into consideration for the interpretation of the results.

RESULTS

The annual food consumption of the different prosimian species has been calculated in terms of fresh weight, and the percentages of consumption of different food categories are shown in Table I. These results present slight differences with previously published data (Hladik, 1978b) which included only a 1-year sample of the same study case.

If the major interspecific differences are obvious (for instance, *Lepilemur* feeding on leaves versus other nonleaf eating species), some other distinctions are more subtle (for instance, the relative proportions of banana and other fruit types eaten by the different species of Cheirogaleinae) and need further analysis in order to demonstrate that they reflect specific behavioral patterns.

The resulting annual mean composition of the diet of the different species is shown in Table II, in terms of percentage of dry weight in protein and fat. Calculations have been done from standard tables complemented by the results of analysis of our samples (especially the G3 Lemur cake; see Hladik, 1978b). No valid comparisons for carbohydrates can be made since these values are generally given by subtraction from the total sample after protein and fat analysis. In fact, considering the marked differences in food choices between foods rich in fiber (leaves), containing large amounts of soluble sugars (apples and pears), or highly concentrated in sugar and

TABLE I

Percentage of the Actual Intake of Different Food Categories in Five Prosimian Species, According to Individual Food Choices from a Standard Set of Foods, over a Period of 2 Years

	Percentage of fresh weight ingested (annual mean)			
Species	Leaves (salad and others)	Apples and pears	Banana	Protein cake (G3) and others
Cheirogaleus medius	0	22	50	27
Microcebus murinus	0	9	42	49
Microcebus coquereli	0	23	42	35
Phaner furcifer	0	13	43	44
Lepilemur ruficaudatus	51	31	0	18

TABLE II

Composition of the Actual Diet of Five Prosimian Species, According to Individual Food Choices from a Standard Set of Foods

Species	Percentage of dry weight (annual mean)	
	Protein	Fat
Cheirogaleus medius	9.2	6.7
Microcebus murinus	11.6	8.0
Microcebus coquereli	9.7	6.5
Phaner furcifer	10.7	7.4
Lepilemur ruficaudatus	12.8	7.2

starch (bananas), a species difference in food composition is likely to be more important for carbohydrates than for other nutrients shown in Table II. In the absence of complete data allowing comparison of the carbohydrate samples, the gross categories of food have been used to indicate specific patterns of food choice. It is noticeable, however, that *Cheirogaleus medius* had the lowest rate of protein and fat intake as a result of a tendency to feed on carbohydrate-rich foods (banana), while the *Lepilemur* species which normally can survive on a very poor diet (Hladik and Charles-Dominique, 1974) presented the highest rate of protein intake during this experiment.

In order to understand the relationships with different physiological cycles observed (Chapter 6) and the meaning of interspecific differences, variations throughout the year of the actual food intake of the different prosimians must be considered. Data presented in Table III have been grouped in four successive "seasons." The actual limits of these seasons were chosen according to the results of the study of specific activity patterns (Chapter 7), at the times where major changes occurred. The records of food consumption during these different periods, obtained in different cages, concerned groups of several animals of a given species. To allow interspecific comparison, the mean value of food intake (fresh weight) per week per 100 grams of body weight have been calculated.

Since many parameters vary simultaneously (intake of various food categories by each species, during different seasons), the multivariable analysis according to the method of Benzecri (1965) (analysis of cor-

TABLE III

Food Intake (in Grams, Fresh Weight, per 100 gm of Body Weight) in Five Nocturnal Prosimians, in Standard Conditions, during Different Periods of the Year (Roughly Corresponding to Seasons)[a]

	December/January/February (Winter)				March/April/May (Spring)				June/July/August (Summer)				September/October/November (Autumn)			
Species	Leaves	Apples and pears	Banana	Protein cake	Leaves	Apples and pears	Banana	Protein cake	Leaves	Apples and pears	Banana	Protein cake	Leaves	Apples and pears	Banana	Protein cake
Cheirogaleus medius	0	13	5 + + +	14	0	12 +	45 +	12 +	0	23	41	28 +	0	3	24	8
Microcebus murinus	0	6	11	17 +	0	6	36 +	30 +	0	6	29	47 +	0	7	38	39 +
Microcebus coquereli	0	27	44	26 +	0	30	65	57 + +	0	25	41	34 + +	0	24	39	43 +
Phaner furcifer	0	11	22	29 +	0	5	36	30 + +	0	6	29	29 + + +	0	11	26	29 + + +
Lepilemur ruficaudatus	41	21	0	17	59	47	0	19 + +	46	30	0	17 +	60	28	0	18 +

[a] Data concerning foods which were totally consumed in the cages during 1 month (+), 2 months (+ +), or 3 months (+ + +) are indicated for each season.

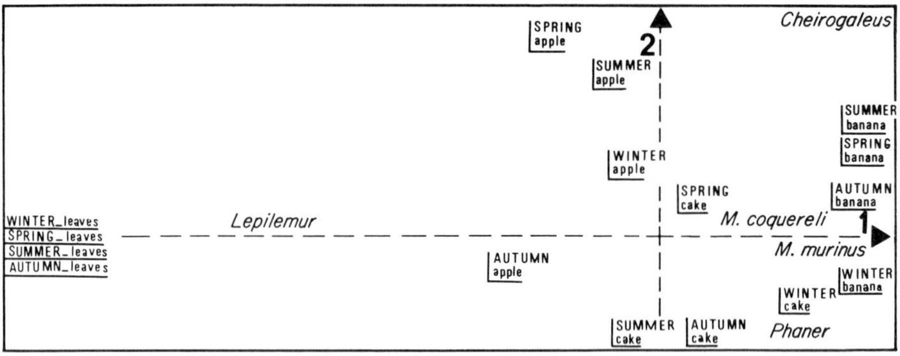

Fig. 1. Multivariable analysis (analysis of correspondences) of food intake in five prosimian species showing interspecific differences as a function of the different food categories eaten during different seasons (projection in the plane of first and second axis).

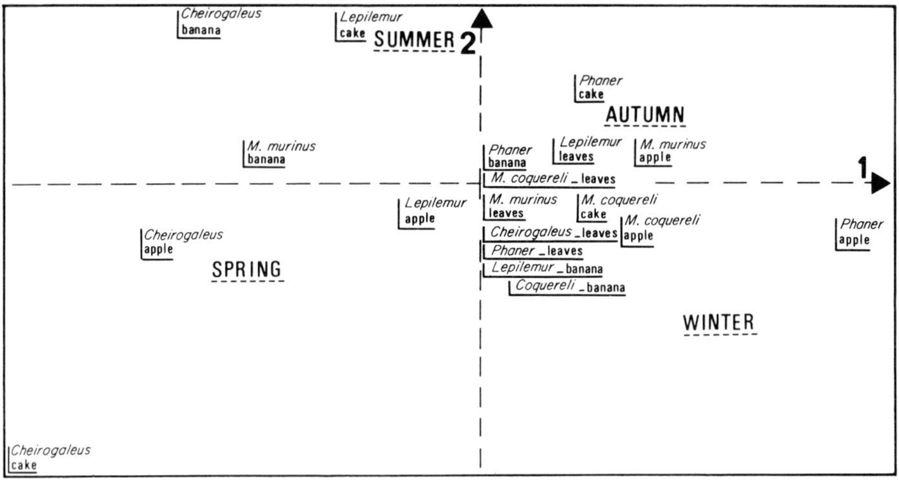

Fig. 2. Multivariable analysis showing seasonal differentiation of food intake of different food categories by five prosimian species in standard conditions.

respondences) was used to show the relative "proximity" or "difference" between species. The results of these analyses* are presented in Figs. 1 and 2. Basic data are those of Table II; however, in order to avoid distortions due to temporary limitation of the available food in some cages, arbitrary values larger than all recorded data have been introduced before data listing.

* We gratefully acknowledge G. Cancela da Fonseca who computed the data at the C.I.R.C.E. (Centre National de la Recherche Scientifique, France) and helped in the interpretation of the results.

These values are:

70 instead of values marked with + in Table III.

85 instead of values marked with + + in Table III.

99 instead of values marked with + + + in Table III.

The specificity of dietary patterns is illustrated in Fig. 1 in which the first axis (1) emphasizes the obvious difference between *Lepilemur ruficaudatus* and all other species. The second axis (2) showing differences of lesser importance in this type of analysis, separates *Cheirogaleus medius,* the two *Microcebus* species, and *Phaner furcifer,* but emphasizes that the feeding behavior of *Microcebus murinus* and *Microcebus coquereli* are closely related in these experimental conditions. The location of the different combinations season/food type illustrates their respective importance in separating the specific feeding patterns.

The relative importance of seasonal variations in these dietary patterns is shown in Fig. 2. During spring and, to some extent during summer, there are marked variations in food intake, translated by dispersion along the first axis (1). Differences between autumn and winter are relatively small as shown along the second axis (2). These variations are not only concerned with the total volume of food intake, but also by qualitative changes illustrated in this diagram by the different combinations species/food type. Seasonal variations, marked in *Cheirogaleus medius, Microcebus murinus* and *Phaner furcifer,* are of lesser importance in *Microcebus coquereli* and *Lepilemur ruficaudatus.*

DISCUSSION

Seasonal variations in body weight and activity of the different prosimian species have been shown to follow the patterns observed in the field, and are mainly synchronized by photoperiodic variations (Chapter 6). Variations in the actual quantity and composition of food eaten by these prosimians also follow the same type of annual cycles, but specific tendencies have strongly affected the overall results. In natural conditions the interspecific differences, in terms of diet, have been interpreted as "adaptive strategies" to different ecological niches (Chapter 2), maintained by particular behavioral patterns which increase the efficiency to obtain "specialized" food resources at different locations in the forest and at different times of the year. In the standard conditions of the present experiment, the maintenance of specific food choices reflects physiological mechanisms which might be the basic support of the behavioral patterns observed in the field.

The interspecific differences illustrated in Fig. 1 (relative "distance" between species) are essentially a qualitative aspect of the combination of different parameters simultaneously considered in this type of analysis. In order to determine to what extent differences are significant (especially in the case of species relatively similar in terms of food choices), the Student t test was applied. All possible pairs of species were compared for monthly intake of each food type and food component (per 100 gm of body weight).

If we designate monthly food intake of each species as x and y, the mean and standard deviation as $\bar{x}$ and $\bar{y}$, S_x *and* S_y, respectively for n_1 number of months x, and n_2 number of months y, then for the hypothesis that the two series are similar:

$$t = (\bar{x} - \bar{y}) \left[\frac{n_1 \, n_2}{(n_1 + n_2) \left[(n_1 - 1) S_X^2 + (n_2 - 1) S_Y^2 / n_1 + n_2 - 2 \right]} \right]^{1/2}$$

According to the number of degrees of freedom, the probability of obtaining the t values actually calculated is extremely low in most cases ($p < 0.01$), especially when the intake of "gross food categories" are compared. This demonstrates that a definite pattern of food intake characterizes each species.

Comparisons of the overall food composition resulting from food choices in artificial conditions yielded a significant difference between *Cheirogaleus medius* and *Microcebus murinus* for protein intake ($p = 0.02$). A similar difference also exists in the natural diet because the smaller species, *Microcebus murinus,* can obtain a relatively larger proportion of insects in addition to the fruits which are available in large quantity (Chapter 2). This purely ecological mechanism is thus maintained and may be accentuated either by a taste specificity or a particular digestive adaptation. Furthermore, the seasonal variation of the digestive ability of *C. medius* may follow the natural seasonal pattern of food availability: nectars of different flowers in November; an increase in available protein (insects) in January/February; and an abundance of fruits in March/April, before the animal hibernates. These seasonal changes in digestive ability of *C. medius* were suggested by Schilling (Chapter 9) who found an increased efficiency of absorption preceding the rest period. However, the variations of quantity and composition of the natural foods may also play a complementary role in the full expression of various physiological cycles which are initially synchronized by photoperiodic variations (see discussion in Chapter 6).

During the annual cycle, the percentage of protein intake in the total diet of *C. medius* and *M. coquereli* varied from 6 to 13, and from 7 to 15, respectively. A similar pattern of annual variation was suggested by the data of Andriantsiferana and Rahandraha (1973) who found that *M. murinus* ate less animal matter during austral summer and spring. In our study, the minimum protein intake occurred at the time of fattening, when the animals ate large amounts of fruit. However, there was a 2-month difference bétween *C. medius* (May/June) and *M. murinus* (July/August), corresponding to the 2-month difference in the respective times of lethargy.

The comparison of the diets of other species in terms of protein and fat content yielded less significant differences (for instance, when comparing *Microcebus murinus* and *M. coquereli*, $p < 0.1$), or were not significant in other cases.

In contrast, the food choices recorded in terms of "gross food categories" permitted definition of particular food patterns for all species, and reflected the preferences for soluble sugars and other types of carbohydrates, which have been discussed in the preceding section. In these cases, the t test demonstrates interspecific differences (which appear on Fig. 1), as well as the constancy in the respective choices of different fruit types in *M. murinus* and *M. coquereli* ($p < 0.01$). The tendency of *M. coquereli* to feed on the pulp of juicy fruits may reflect the propensity of this species to seek different natural sweet secretions and exudates (see Chapter 4).

In *Phaner furcifer*, a partly similar tendency was found in autumn/winter (Table III). Interpretation of this food choice is more delicate in this case, since the gum on which the species feed in nature (Chapter 3) was totally absent in captivity, and no type of carbohydrate among those provided was of similar composition. Nevertheless, the slight seasonal change in diet accompanied by a slight variation in body weight and activity (Chapter 6) may also reflect a periodical change in dietary physiology at the time normally requiring maximum specialization (on gums) in natural conditions.

The same remarks also apply to *Lepilemur ruficaudatus* in which the tendency to eat more fruit was accentuated in spring, while the maximum intake of leaves was in autumn. In field conditions, this would occur at the time of leaf flushing of most species. Young leaves have a maximum protein content (Hladik, 1978a) which explains the apparently surprising result for *Lepilemur*, in which the diet includes more protein than that of other species.

The mechanisms of these seasonal variations and specific patterns of food choices are still hypothetical. It was suggested (Hladik *et al.*, 1971; Hladik, 1977) that the relative balance between long-term conditioning to food intake (after absorption of nutrients) and the immediate response (taste and motivation) would result in different patterns of food choices. This hypothesis may be sufficient to explain why some folivorous primates feed on the most common and least nutritious plants, whereas other sympatric primate species feed on a variety of richer foods (leaf flush and fruits) which are more scattered in the forest (Hladik, 1977). The first pattern would mainly be based on long-term responses, while the species showing the second pattern would be motivated to look for scattered sources of carbohydrates and fat, which not only satisfy hunger, but also provide sufficient reward, in terms of taste pleasure, to maintain a high motivation.

Morphological adaptations of the digestive tract are important in both cases and have been discussed elsewhere (Chivers and Hladik, 1980). Nevertheless, motivation and possibly intensity of taste sensation can result in major differences in specific food choices.

The specific patterns of food choices demonstrated by the nocturnal prosimians of the West Coast of Madagascar can also be explained by the relative importance of taste stimulation and motivation, which may fluctuate during the course of the annual cycle, as suggested by the results concerning *Lepilemur ruficaudatus* and *Phaner furcifer*. Nevertheless, highly specific patterns, such as the tendency of *Microcebus coquereli* to find sweet liquid food, or the maintenance of a protein/carbohydrate balance similar to that observed in the field for *Microcebus murinus* and *Cheirogaleus medius,* may be regulated by as yet unknown mechanisms.

The significant finding of this study is the relationship shown between the more or less accentuated seasonal variations in specific food choices, and the seasonal availability of food resources, which constitutes an adaptation to a highly variable environment.

REFERENCES

Adriantsiferana, R., and Rahandraha, T. (1973). Variation saisonnière du choix alimentaire spontané chez *Microcebus murinus. C.R. Hebd. Seances Acad. Sci. Ser. D 277,* 2025–2028.

Benzecri, J. P. (1965). "Sur l'Analyse Factorielle des Proximitées." I.S.U.P., Paris.

Chivers, D.J., and Hladik, C.M. (1980). Morphology of the gastrointestinal tract in Primates: Comparisons with other mammals in relation to diet. (in press).

Hladik, C.M. (1977). A comparative study of the feeding strategies of two sympatric species of leaf monkeys: *Presbytis senex* and *Presbytis entellus*. *In* "Primate Ecology" (T.H. Clutton-Brock, ed.), pp. 323–353. Academic Press, New York.

Hladik, C.M. (1978a). Adaptative strategies of Primates in relation to leaf-eating. *In* "The Ecology of Arborel Folivores" (G.G. Montgomery, ed.), pp. 373–395. Smithson. Inst. Press, Washington, D.C.

Hladik, C.M. (1978b). Diet and ecology of prosimians. *In* "The Study of Prosimian Behavior" (G.A. Doyle and R.D. Martin, eds.), pp. 307–357. Academic Press, New York.

Hladik, C.M., and Charles-Dominique, P. (1974). The behaviour and ecology of the sportive lemur (*Lepilemur mustelinus*) in relation to its dietary peculiarities. *In* "Prosimian Biology" (R.D. Martin, G.A. Doyle, and A.C. Walker, eds.), pp. 23–37. Duckworth, London.

Hladik, C.M., Hladik, A., Bousset, J., Valdebouze, P., Viroben, G., and Delort-Laval, J. (1971). Le régime alimentaire des Primates de L'Ile de Barro Colorado (Panama): Résultats des analyses quantitatives. *Folia Primatol.* **16,** 85–122.

Petter, J.J. (1978). Ecological and physiological adaptations of five sympatric nocturnal lemurs to seasonal variations in food production. *In* "Recent Advances in Primatology" (D.J. Chivers and J. Herbert, eds.), Vol. 1, pp. 211–223. Academic Press, New York.

9 *Seasonal Variation in the Fecal Marking of Cheirogaleus medius in Simulated Climatic Conditions*

A. SCHILLING

INTRODUCTION

Cheirogaleus medius is a nocturnal prosimian commonly found in all arid areas of Madagascar (Petter *et al.*, 1977). Other nocturnal prosimians, such as *Lepilemur* and *Phaner,* have adopted particular strategies to cope with the periodically severe environmental conditions (Hladik and Charles-Dominique, 1974; Chapter 3). The different species of *Cheirogaleus* adjust to seasonal variations in an original manner both as regards behavior (economy of movements), and physiology (*Cheirogaleus* species together with *Microcebus murinus* are the only primates to show seasonal lethargy (Chevillard, 1976; Chapter 6). In our laboratory, climatic conditions recorded in the field were reproduced but failed to bring about the winter lethargy which lasts about 6 months under natural conditions (Petter, 1978). However a noticeable reduction of activity was obtained in the winter semester, during which the animals rarely and irregularly left the nest box (Chapter 6).

An additional peculiarity distinguishing the genus *Cheirogaleus* from other prosimian and primate species is the regular use of the feces as a scent marking material (Schilling, 1978). Indeed, the protruded anus of *Cheirogaleus* (Hill, 1953) constitutes a remarkably efficient marking organ. The fecal marking behavior, which has been observed in the field (Petter, 1962, and personal observation), consists of a dragging defecation occurring during locomotion, resulting in elongated cylindrical marks (about 10 cm in length in *C. medius,* 40 cm in *C. major*) more or less flattened as the animals wipe or trample them. This is often accompanied by urination.

In captivity, the marks are obviously more concentrated, forming a thick

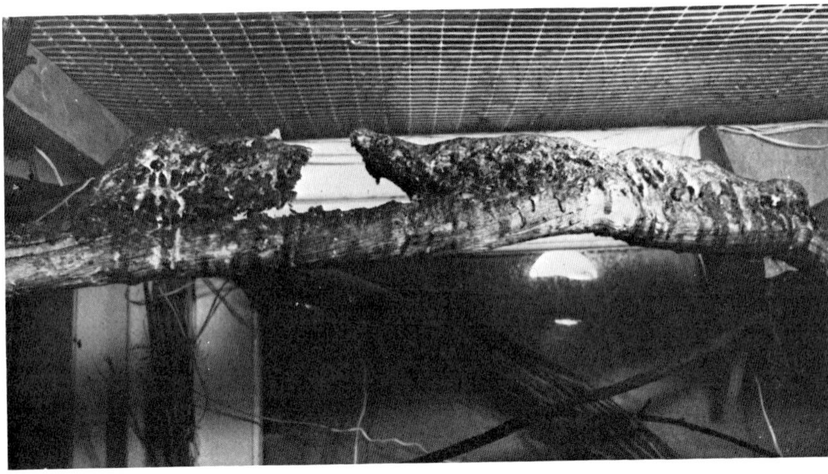

Fig. 1. Fecal mark accumulation (about 5 cm thick) at a place of frequent marking of *C. medius* in the animal house.

layer of dried feces on certain branches or supports (Fig. 1). In *C. medius* this particular behavior appears to be related to the seasonal cycle of activity. Preliminary observations showed that during the active semester fecal marking increased and, moreover, a noticeable peak occurred at a precise moment within this period. It was thus attempted to study the variation of this marking in relation to variations of locomotor activity and food consumption, in order to determine to what extent fecal marking is correlated with these parameters. If fecal marking varies independently, this would provide evidence that this type of defecation is a particular behavior as in canids, lagomorphs, ruminants, rodents, and insectivores, and may be more than the mere result of physiological functions.

METHODS

Experimental Conditions

An adult male and female pair of *Cheirogaleus medius* obtained from Madagascar in January 1973 was used in this study. The photoperiodic, temperature, and food conditions in captivity are described elsewhere (Chapter 6). Recording was done over a period including more than two annual cycles, from April 15, 1974, to October 15, 1976. The cage occupied by this small-sized animal (body length about 20 cm) measured 1.90 × 2.0 × 2.2 m, was installed with branches and lianas at the beginning of the experiment, and was isolated by black curtains. Disturbances were minimized and limited to change of food and checking of the activity recording equipment.

Recording of Activity

Locomotor activity was recorded by six platforms distributed in the cage and equipped with microswitches connected to numerical counters which were checked daily (see Chapter 7 for details).

The monthly average of daily records from June 1974 until October 15, 1976 was used to represent the graphic plot of the activity or "actogram."

Measure of Food Intake

The amount of the different categories of food consumed on a regular diet both in quantity and quality were noted every day (see Chapter 8). The following values were calculated daily either by drying in an oven or by using standard tables (Experientia-Geigy): (1) The dry weight of food consumed; and (2) the weight of nonassimilable food consumed, mainly fiber of fruits. This last datum, calculated according to standard values for humans, should reflect more accurately the basic part of the feces, as long as we take the liberty of comparing the digestive physiology of two rather different primates!

Measure of Fecal Marking Activity

Filter papers were placed at the same location on branches, and were changed every month over a period of 30 months (from April 15, 1974, to October 15, 1976), to allow estimation of the variation in the amount of feces deposited by the animals throughout the yearly cycle. The filters were dehydrated in an oven (60°C) and weighed before being placed in the cage. At the end of each month the papers with deposited feces were also dehydrated and weighed. The difference was used as a "sample mark" and represented only a part of the total amount of feces in the cage. Since preliminary observations in 1973 indicated that the amount of feces deposited by *Cheirogaleus* showed a sharp peak around the first of August, the feces were weighed the 15th day of each month so as to better locate the annual peaks.

RESULTS

Marking and Locomotor Activity

In consideration of the results of the actogram, the annual activity cycle has been divided into an active semester and an "inactive" semester which corresponds to the animal's period of lethargy in Madagascar (Fig. 2). It should be noted that under our experimental conditions which imperfectly reproduced the natural climate, not only did a residual activity persist during the "inactive" semester, but the active period (from March 15 to September 15) was 1 month in advance of that expected from field data (Fig. 2). Comparing the data obtained over a period of 2 years, it may be noticed that although 83 ± 4% of the locomotor activity and 85 ± 4% of the marks took place during the active semester, the two parameters fluctuated independently (Fig. 3). Actually the peak of the actogram which was situated at the beginning of the summer (44.3% of the annual locomotor activity between May 15 and July 15) corresponds to a relative pause of the increase in the amount of sample marks. Indeed, the peak of this fecal marking occurred later during this period (45.2% of the annual marking between July 15 and September 15), while locomotor activity decreased.

In addition, the annual peak of fecal marks has decreased and been successively retarded during the 3-year period of observation (beginning of August 1974, middle of August 1975, beginning of September 1976, Fig. 2) possibly due to the effects of captivity.

Marking and Food Consumption

These two parameters are obviously related. Thus both the annual peak of dry weight of sample marks as well as that of food or "fibers" consumed, was attained between July and September 15 (Figs. 2 and 3). However, during the inactive semester, these two measures evolved independently: marking activity was very low (15% of the yearly amount) despite substantial feeding activity (34%). Similarly, the active semester was characterized by two periods during which marking and feeding were less closely related than during the remainder of this semester.

The first period took place from April 15 to July 15, during which the curve of the dry weight of the sample marks followed more closely the dry weight of the nonassimilable part (fibers) of the food, rather than that of the dry weight of food consumed (Fig. 2). Variations in the average food composition depend on the seasonal variations in food choices (see Chapter 8, Table I). However, at the time of the spring peak of fecal marks (May 15–June 15 in 1975, April 15–May 15 in 1976) if the percentage of the sam-

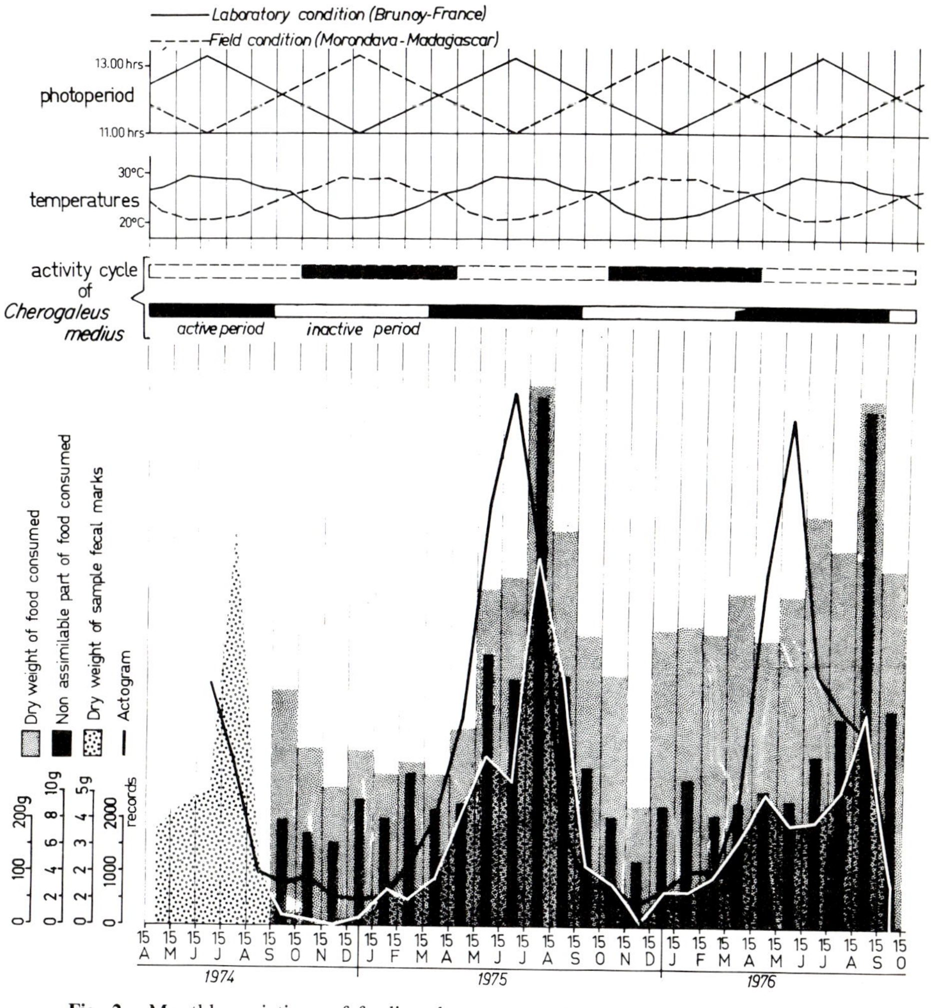

Fig. 2. Monthly variations of feeding, locomotor activity (black line), and fecal marks (white line) of *Cheirogaleus medius,* in relation to simulated climatic conditions.

ple marks compared to the annual total remained constant (12.7% in 1975, 13.2% in 1976) the weight of the "fibers" of food consumed over the same period showed a large variation (respectively, 12.4% in 1975 and 6.9% in 1976).

The second period (September 15–October 15) during which the weight of the marks appeared less related to food consumption, immediately followed the annual peak in marking behavior. Food consumption re-

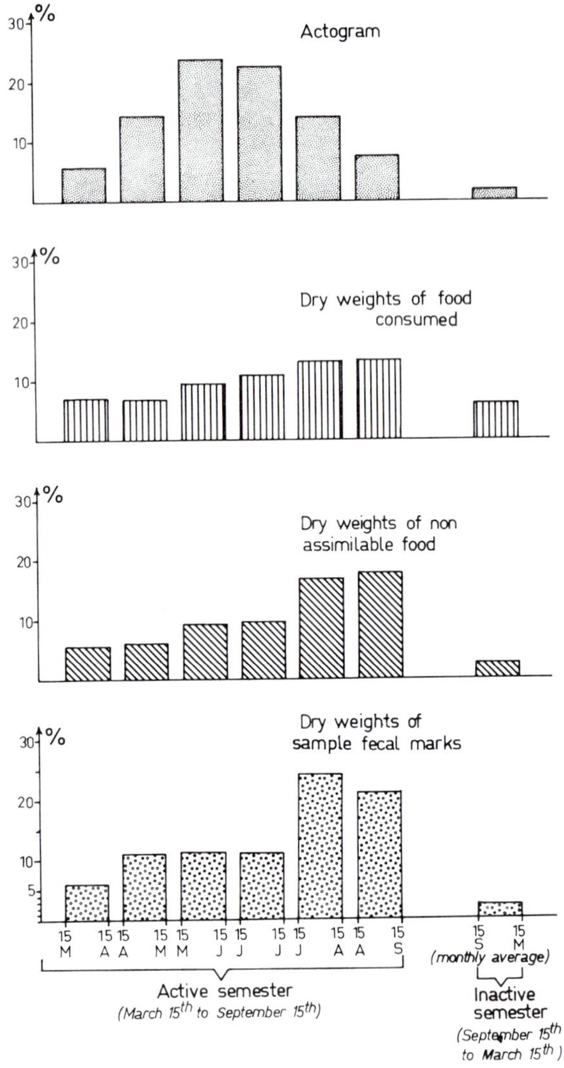

Fig. 3. Monthly percentages of the annual values of activity (according to number of records), feeding (in terms of total food consumed and amount of nonassimilable part), and fecal marks of *Cheirogaleus medius*. The residual activity during the inactive semester (monthly average value) is probably smaller in natural conditions.

mained high (9.5% of the annual dry weight of food and 9% of the fibers) whereas the marks decreased to 5% of the annual total.

Significance of Marking in Relation to Locomotor and Feeding Activities

During certain periods of the seasonal activity cycle of *Cheirogaleus medius,* marking behavior seemed to evolve independently of locomotor activity and food consumption. To make these comparisons clearer the following ratios were calculated which will be referred to as marking indices (Im):

Im_1 = the ratio of the dry weight of the sample fecal marks to the locomotor activity

Im_2 = the ratio of the dry weight of the sample fecal marks to the dry weight of the food consumed

Im_3 = the ratio of the dry weight of the sample fecal marks to the weight of the nonassimilable fraction of consumed food

The seasonal variations of these indices are presented in Fig. 4. The fecal marking indices followed the same pattern only during the inactive period (September 15–March 15). Moreover, during the active period several different phases may be defined according to these indices:

In spring, a phase characterized by the steady decrease of the marking index Im_1, and an isolated peak of the marking indices Im_2 and Im_3 (spring peak) resulted in a clear inversion of the annual curves of these indices (Im_2 or $Im_3 > Im_1$) in March/April. This phase during which the marking activity is comparatively high in relation to food consumption and low in relation to locomotor activity occurred at the time of the mating period under natural conditions (Petter, 1978).

At the beginning of summer (May 15–July 15), there was a decline of all the indices resulting from a relative pause in fecal marking compared to the other activities.

At the end of the summer (from about July 15 to the beginning of the "inactive" semester, September 15) a phase occurred during which the marking activity increased in relation to food consumption, and even more so in relation to locomotor activity. This phase is consequently characterized by an annual peak for the three indices which all increased together during summer but declined with a noticeable delay for Im_1 and also with an inversion of the ratios ($Im_1 > Im_2$ or Im_3). This phase, during which the fecal marking activity was fairly important in relation to locomotor activity but small in relation to food consumption, occurred at the time of the annual fattening of the animals preceding the inactive period (winter lethargy, Chapter 6).

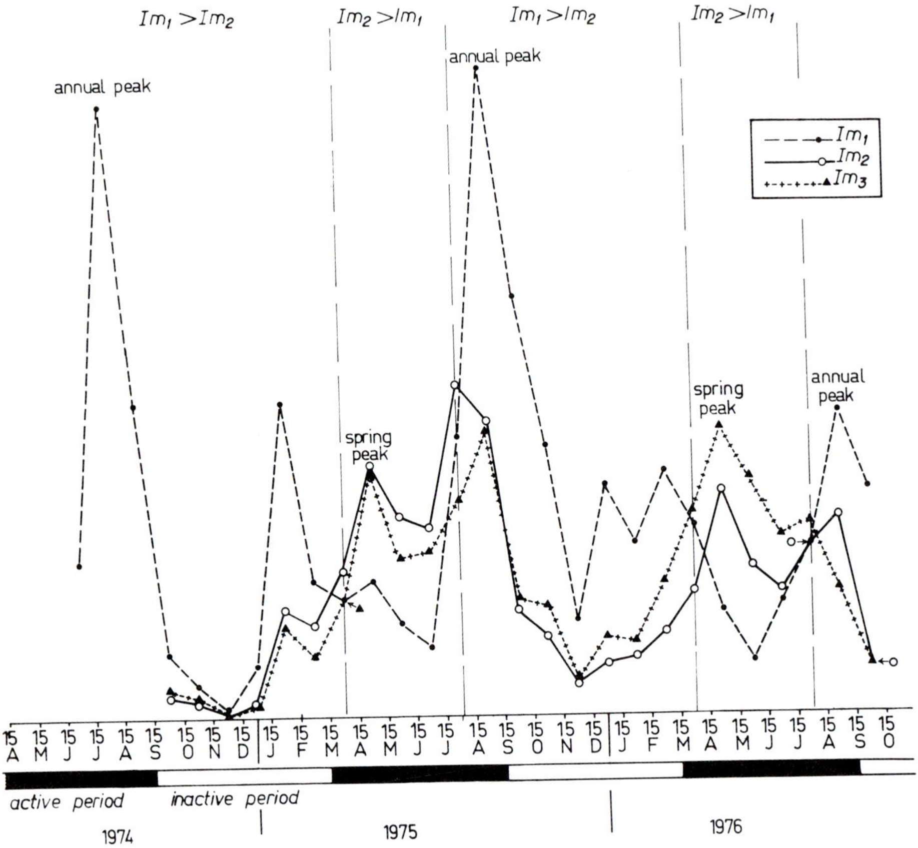

Fig. 4. Seasonal variations of the fecal marking indices of *Cheirogaleus medius*—Im_1 = fecal marks/locomotor activity, Im_2 = fecal marks/total food intake, Im_3 = fecal marks/fiber intake.

DISCUSSION

The results show that fecal deposition in *Cheirogaleus medius* is complex, due to the seasonal variations, and implies a particular physiological function. If this fecal marking behavior is clearly independent from locomotor activity, further experimentation is needed to explain the apparent variability with respect to food consumption and to provide a possible explanation of the two characteristic peaks which occur during the active semester. A possible explanation for these peaks may result from seasonal variations in topography of feces deposition. For example, animals may have defecated more on the branches used for sampling, during May (spring peak) and August (annual peak) and subsequently this would have resulted in a bias in the marking indices. However, experiments in progress demonstrate that, although branches are marked differentially, these differences remain constant throughout the year. Furthermore the quantity of feces that falls to the ground is insignificant in relation to that on the branches.

A second possible explanation concerns nutritional physiology. It may be assumed that in *Cheirogaleus* a seasonal variation exists in food assimilation in relation to annual lethargy. The spring peaks of Im_2 and Im_3 would thus result from a decrease in the ratio food assimilated/food consumed. At this time (May) the animals, food intake increased while body weight decreased (see Chapter 6, Fig. 2). The increase in food intake (by a factor of 2) with a corresponding decrease in weight (10%) cannot be explained only by a difference in activity (5% in relation to annual mean, Chapter 6). During this period, at least in natural conditions, the animals resume territorial and sexual activities (Chapter 2). The annual peak of fecal marking may be explained in a similar manner. However whereas the coefficient of assimilation is high, the animal tends to consume an excess of food resulting in a corresponding increase of fecal material in August/September.

Food intake decreased in September but was still sufficient for fat accumulation (high coefficient of assimilation), the seasonal peak in body weight occurred about 1 month after the peak in food intake and was immediately followed by the period of lethargy. This particular physiology, as yet insufficiently understood, results in seasonal variations in the amount of the fecal material.

The use of the feces as a marking material can only be demonstrated according to additional behavioral observations. As for urine washing in other prosimians, fecal marking in *Cheirogaleus* is a stereotyped behavior which appears as early as 6 weeks of age (Schilling, 1978). The same type of anal dragging has also been observed in the absence of fecal deposition. These marks are sniffed by conspecifics of both sexes. Finally, laboratory observations on *C. major* show that fecal marking tends to be performed at the periphery of the available space. This again resembles urine washing in *Galago* which occurs at the periphery of the home range both in the laboratory and in the field (Charles-Dominique, 1974, 1977a,b).

Cheirogaleus fecal marks have indeed been observed in nature (Petter, 1962), although no information concerning their role in territoriality is available. Since scent marks can be discriminated after several months in certain conditions (Schilling, 1978), fecal marks may also be very durable, especially during the dry season. If the large quantity of marks deposited before hibernation play a role 6 months later in the reestablishment of home ranges (see Chapter 2) then the physiological variations discussed in this paper would be complementary to the behavioral function of marking.

REFERENCES

Charles-Dominique, P. (1974). Ecology and feeding behaviour of five sympatric Lorisids in Gabon. *In* "Prosimian Biology" (R. D. Martin, G. A. Doyle, and A. C. Walker, eds.), pp. 131–150. Duckworth, London.

Charles-Dominique, P. (1977a). "Ecology and Behaviour of Nocturnal Primates." Duckworth, London.

Charles-Dominique, P. (1977b). Urine marking and territoriality in *Galago alleni.* A field study by radio-telemetry. *Z. Tierspsychol.* **43,** 113–138.

Chevillard, M. C. (1976). Capacités thermorégulatrices d'un Lémurien malgache, *Microcebus murinus* (Miller 1777). Thesis (3rd cycle), University of Paris VII.

Hill, W. C. O. (1953). "Primate: Comparative Anatomy and Taxonomy," Vol. I. Edinburgh Univ. Press, Edinburgh.

Hladik, C. M., and Charles-Dominique, P. (1974). The sportive Lemur. In "Prosimian Biology" (R. D. Martin, G. A. Doyle, and A. C. Walker, eds.), pp. 23–37. Duckworth, London.

Petter, J. J. (1962). Recherche sur l'écologie et l'éthologie des Lémuriens malgaches. *Mem. Mus. Natl. Hist. Nat., Ser. A* **27,** 1-146.

Petter, J. J. (1978). Ecological and physiological adaptation of five sympatric Lemurs to seasonal variations in food production. *In* "Recent Advances in Primatology" (D. J. Chivers and J. Herbert, eds.), Vol. 1, pp. 211–223. Academic Press, New York.

Petter, J. J., Albignac, R., and Rumpler, Y. (1977). Mammifères Lémuriens (Primates Prosimiens). *Faune Madagascar* **44.**

Schilling, A. (1980). The possible role of urine in territoriality of some nocturnal prosimians. *Symp. Zool. Soc. London* (1980).

10 *Ecological Correlates of Visual Learning in Nocturnal Prosimians*

H. M. COOPER

INTRODUCTION

The study of learning from a comparative viewpoint poses numerous theoretical and practical problems (Warren, 1973). Of fundamental concern are the criteria for deciding the type of learning to be studied and the theoretical bases for comparison of the species concerned. Among the various approaches, the main trends have been to compare learning abilities either to measures of cortical development (Masterton *et al.*, 1974; Riddell *et al.*, 1974, 1976), phylogenetic position (Bitterman, 1965; Rumbaugh and Gill, 1972), or to species-specific behavioral capacities (Bolles, 1970; Seligman, 1970).

The present approach is based on the analysis of prosimian learning in relation to ecological and behavioral characters discussed in the previous chapters of this volume. Learning was considered as an integral part of the various species adaptations to the environment, and thus the main emphasis has been focused on the functional roles of learning capacities. This type of functional approach has often been used to analyze the significance of similarities and differences observed for many biological traits (Kay, 1975; Szalay and Seligsohn, 1977).

However, whereas certain characters (such as dental or skeletal) can be directly measured and described, and their functional role easily assessed from behavioral observations, the role of learning in the development and maintenance of behavioral patterns is more difficult to define. A means of assessing these capacities has been to conduct tests under controlled laboratory conditions. In terms of complex learning processes, such tests are merely a distorted reflection of these capacities, since they are necessarily removed from the environmental context in which they normally develop and function. Nevertheless, these procedures provide an experimental situation in which the various parameters can be systematically controlled, the results of which must be subsequently confronted with field observations.

Two species of nocturnal prosimians, *Microcebus murinus* and *Phaner furcifer,* were used in the present behavioral study. Comparisons between these two species are facilitated by the fact that they share recent common ancestry (Charles-Dominique and Martin, 1970) but are well differentiated according to several ecological and anatomical characters (Chapters 2 and 3). In order to be able to relate these distinctions to visual learning capacities, a number of behavioral measures were employed (spatial and visual reversals, object discrimination) which require different types of abilities for what is normally referred to as "problem solution." Since the quantitative results of a single type of test are difficult to compare between species due to differences in perception and motivation (Warren, 1973), comparisons were made by using combined results of a number of tests to establish initially a "species profile." In this investigation, the profile allows a qualitative appreciation which takes into account capacities for attention and responsiveness to different stimulus cues, discrimination abilities, flexibility of learning, and memory. The profile provides a basis for comparison between species, and a means for relating learning to other features of the species biology.

MATERIAL AND METHODS

Subjects

The animals used in this study included eight *Microcebus murinus* and two *Phaner furcifer*. All were adult animals that had been maintained for several years in captivity. *Phaner* were housed individually in cages which measured 2.0 × 1.5 × 1.0 m, while *Microcebus* were kept either individually or as pairs in cages 1.0 × 0.5 × 0.5 m. Cages were constructed of wood and wire mesh in which visual communication between cages was prevented. For each species, sliding trap doors in the cage gave access to a central corridor which led to the test chamber, enabling each individual to be tested separately. The daily light cycle was inversed to allow testing during the species active period. Animals were maintained on a varied diet of fruits, milk, and supplementary cereal; water was available at all times. Prior to testing, animals were food-deprived for 18–20 hours. Food rewards depended on species and individual preferences, and consisted of life mealworms or sweetened milk.

Apparatus

The apparatus was designed after that of Rumbaugh *et al.* (1972) with the additional possibility of automatically presenting three-dimensional objects as stimuli. The stimulus windows measured 6.0 × 6.0 cm and were located 2.5 cm apart on the front panel of the apparatus. The stimulus windows consisted of a one-way mirror, behind which were located the stimulus objects. The stimuli were visible only when the overhead stimulus lights were illuminated. The response was detected when the subject interrupted an infrared-photocell circuit located directly in front of each stimulus window. A 5.0 cm long wire mesh screen separated the two windows but allowed access to the food well situated between them. A starting platform was located at the rear of the test chamber.

The programming of stimulus position, reward, performance criteria, and recording of results were controlled by electromechanical and electronic systems located in a separate room, from which the experimenter could also observe the animals' behavior via infrared video camera equipment.

Procedure and Testing

Shaping of the response was achieved in several stages, but essentially involved initially training the subject to break the photocell light beam for reward, to respond to both stimulus windows, and eventually to acquire an observing response from the starting platform.

During the actual test situation, the onset of stimulus presentation was triggered by the animal's weight on the starting platform. The stimulus lights remained on until a response was made by the subject. Any response immediately resulted in the extinction of the stimulus lights and thus the disappearance of the stimuli. Only correct responses were rewarded and the noncorrection procedure was employed. A 5-second intertrial interval followed, subsequent to which the subject was again required to activate the starting platform to initiate a new trial. A total of no more than 60 responses was allowed during an experimental session which usually lasted 20 minutes.

Three types of discrimination problems were given.

1. *Successive visual discrimination reversals (visual SDR):* The relevant stimuli consisted of black and white cards, which changed position randomly. One of these stimulus cards was designated as positive and associated with the food reward; all spatial cues were irrelevant to problem solution. The subject was required to achieve a criterion of 10 correct

responses in 10 consecutive trials, after which the relative values of the stimuli were reversed. A total of 15 reversals were given, followed by a supplementary reversal and a nonreversal each after intervals of 30 days.
2. *Successive spatial discrimination reversals (spatial SDR):* The same black and white stimulus cards, as above, were present in the stimulus windows and changed sides randomly, but were unrelated to reward (nonspatial irrelevant visual cues). Reward was associated with the spatial aspects of the test situation, and the subject was required to respond to the stimulus window designated as positive. The same performance criterion as above was used. A total of 50 reversals were given, however during reversals 31–40 the irrelevant visual cues were replaced by equivalent gray cards.
3. *Transfer Index (TI):* As an additional measure of visual discrimination learning ability the Transfer Index developed by Rumbaugh (1969) was used. Each problem consists of a pair of three-dimensional stimulus objects. The subject is required to attain an acquisition criterion of either 67 or 84% correct responses, after which cue values are reversed during a limited number of trials (10). The TI is the ratio of percentage correct responses during reversal, to the percentage correct responses during acquisition for ten problems. For details and the logic of this method, consult Rumbaugh (1969). During TI testing an additional modification was added to form a distinct observing response. The intertrial interval was variable (5–15 seconds) and the subject had to remain on the starting platform for 1.5 seconds after the onset of the stimulus lights. If the subject left the platform before completion of the observing response, the lights were extinguished and the intertrial interval reinitiated. This method required the subject to maintain a high level of attention to the visual stimuli.

Only one reversal, or a single TI problem, was given per day. The positions of the stimuli for all discriminations were determined by the Gellermann (1933) series.

Three *Microcebus* were used in the spatial SDR, and three in the visual SDR. Two of the latter received further TI problems as well as two additional naive individuals. Of the two *Phaner* available, one had spatial SDR followed by visual SDR, the other individual had visual SDR followed by spatial SDR, and both subsequently received TI problems.

RESULTS

Visual SDR

The results of 15 visual reversals for both species are shown in Fig. 1. On the initial visual discrimination (0) the mean number of errors made by

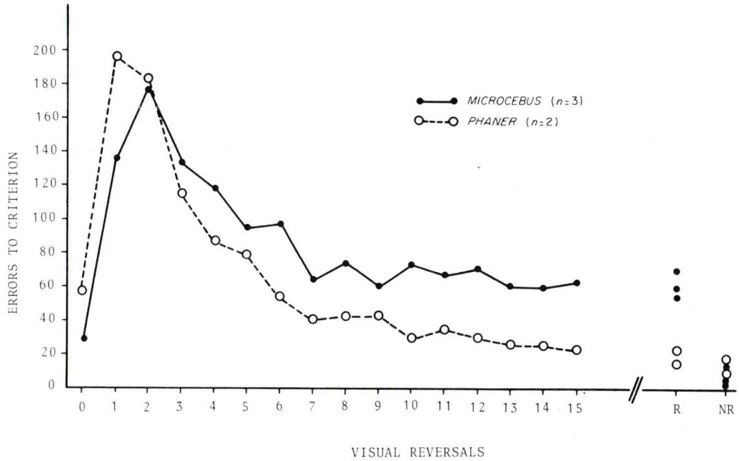

Fig. 1. Number of errors to reach criterion in *Microcebus murinus* and *Phaner furcifer* for 15 successive visual discrimination reversals (group curves), and for additional reversal (R) and nonreversal (NR) after 30-day intervals (individual scores).

Microcebus is lower than that of *Phaner*. This relation changes however over the subsequent series of reversals. Both species made a large number of errors on the initial reversals but eventually demonstrated progressive improvement. The improvement in performance of *Phaner* is more important than that of *Microcebus* especially during later reversals. *Phaner* eventually solved later reversals with fewer errors than the original discrimination, whereas this was not the case for *Microcebus*.

In both species the large number of errors made on the initial reversals is associated with a tendency to persevere in responding to the formerly positive stimulus after each reversal. The intraproblem learning curves (Fig. 2) show that after each reversal, the animals always began responding below chance (50%). This tendency became progressively reduced in *Phaner,* but persisted even throughout later reversals in *Microcebus*.

Since this result suggested a difference in capacities for memory and/or forgetting, an additional reversal (R) and a nonreversal (NR) were given, each after intervals of 30 days (in random order for each individual, Fig. 1). *Phaner* showed no difference in the number of errors under either condition, and made as many errors as on the previous reversal. On reversal R, *Microcebus* also made as many errors as on the previous reversal, however, almost no errors on the nonreversal. This indicates that whereas for *Phaner* forgetting has occurred during this period of time, no forgetting has been demonstrated by *Microcebus*.

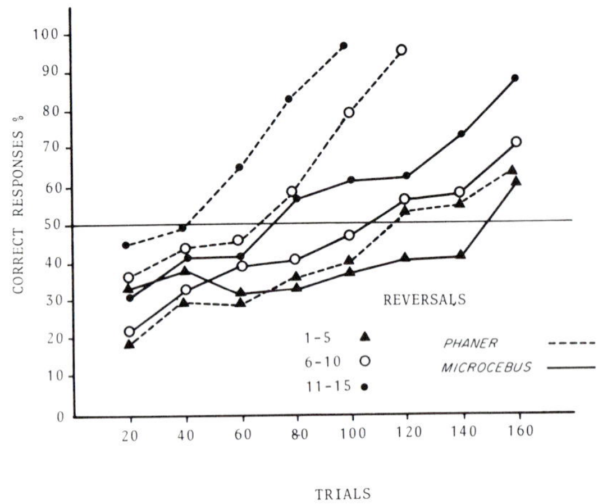

Fig. 2. Intraproblem learning curves for visual discriminations in *Microcebus murinus* and *Phaner furcifer*.

Spatial SDR

Performance of *Phaner* and *Microcebus* on the spatial SDR with irrelevant cues present, resulted in an inversion of the learning curves as compared to the previous visual SDR. Figure 3 shows the mean number of errors to reach the criterion for blocks of 10 successive reversals. As in the visual reversals, both species demonstrate progressive improvement, but in this case, *Microcebus* makes fewer errors than *Phaner* when irrelevant nonspatial visual cues are present. When the black and white stimulus cards were absent (replaced by equivalent gray cards during reversal 31–40), the performance of both species was similar, and few errors were made. However, at this point of training both species had sufficient learning experience and made few errors. In contrast, the reintroduction of irrelevant visual cues (reversals 41–50) produced a disruption in the performance of *Phaner* but not in that of *Microcebus*. Thus the presence of irrelevant visual information produced a more distracting effect only on the former species.

Transfer Index

The TI is a useful measure of learning behavior since it provides not only an indication of discrimination ability during acquisition, but also reflects behavioral flexibility during reversal. In addition the TI values permit comparison between species equated for similar acquisition before reversal.

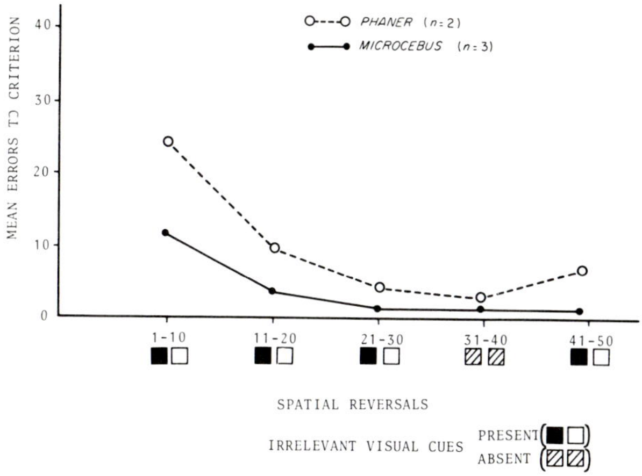

Fig. 3. Mean number of errors to reach criterion for successive spatial discrimination reversals with irrelevant visual cues present or absent.

As can be seen from Table I, both species are capable of relatively rapid solution of discrimination problems, as compared to other primates (Rumbaugh and Gill, 1972). At both the 67 and 84% acquisition levels *Phaner* solved object discriminations in fewer trials than *Microcebus*. *Microcebus* required a mean of 8.2 more trials to attain the 67% criterion level, and 8.7 more trials than *Phaner* to attain the 87% level. However, there is no significant difference between species in the percentage of correct responses during reversal at either level. Thus, although the acquisition performance of *Phaner* is superior to that of *Microcebus* this difference in learning is not

TABLE I

Transfer Index Scores for *Microcebus murinus* and *Phaner furcifer*

Species	Criterion acquisition (% correct)	Total number of problems	Average number of trials to reach criterion	Reversal (% correct)	Transfer index (reversal % / acquisition %)
Microcebus	67	60	28.6	41.5	0.62
murinus	84	60	40.1	22.8	0.27
$n = 4$					
Phaner	67	20	20.4	44.6	0.67
furcifer	84	40	31.4	23.1	0.28
$n = 2$					

reflected during reversal and, accordingly, not in the TI values. Both species show negative transfer of learning (i.e., TI < 100; Rumbaugh, 1969), which is relatively more important after 84% acquisition than after 67% acquisition. Indeed the reversal performance merely reflects the fact that each species continues to respond to the initially positive stimulus in the same proportion during reversal as in the previous phase of acquisition. Thus when the acquisition criterion is raised from 67 to 84% correct responses (17% difference) reversal performance drops proportionately (18.7% for *Microcebus,* 21.5% for *Phaner*).

Discrimination Behavior

Observation of the animal's behavior during testing is indispensable for understanding the importance of different stimuli, and their significance for the species during discrimination. In discrimination problems such as those given in this study, certain cues are relevant to problem solution, while others are irrelevant. The behavior of various species, which perceive or respond differentially to these cues, will be qualitatively different. In fact, the entire test situation provides cues which may influence the final quantitative results.

In the present study, *Microcebus* and *Phaner* differed in behavior on both visual and spatial problems. During all types of testing, *Microcebus murinus* made its choice initially from the starting platform. However, during the spatial SDR this species paid attention to the global aspects of the test chamber rather than to the stimulus windows. When the onset of a trial was detected by the illumination of the stimulus lights, *Microcebus* approached the food well by following one of the walls of the apparatus, easily ignoring the irrelevant nonspatial visual cues (black and white cards) in the stimuli windows. In contrast, during visual problems, in which these same cues were relevant and rewarded, *Microcebus* strongly focused its attention toward the stimulus windows from a distance before making a rapid approach and immediate response.

Phaner always approached the stimulus windows by running directly to the front of the apparatus, and inspected both windows before making a response. *Phaner* thus had great difficulty in ignoring the irrelevant nonspatial visual cues during spatial SDR. This was especially evident on early reversals and also after reintroduction of the irrelevant stimuli (reversals 41–50). During the visual reversals, behavior was essentially similar but included more careful inspection of the stimuli before response.

Thus depending on the situation, both species focus their attention differentially to the stimulus cues. The choice in *Microcebus,* especially for visual problems, is based on scanning and decision from a distance, whereas discrimination in *Phaner* is typified by inspection at close range.

DISCUSSION

According to the results of the present study, visual learning and associated behaviors are clearly different in the prosimians *Microcebus* and *Phaner*. Both species learn spatial and visual problems but with relatively different degrees of difficulty. This difference is related to the way in which the animals attend to various aspects of the environment, and the integration and subsequent utilization of visual information.

In *Microcebus* spatial reversals of a discrimination problem are learned comparatively more rapidly than visual reversals, since this species appears to be more prepared to attend to the spatial aspects of the present test situation. Spatial cues do not exclude the use of visual cues, but may also include other types of sensory information. However, the fact that individual object discriminations are rapidly acquired reflects a significant capacity for visual attention and learning. Thus, when nonspatial visual cues are relevant, *Microcebus* can ignore the spatial cues which, when rewarded, are highly preferred. This tendency to attend to the visual stimuli is also apparent from the perseverance to the formerly rewarded stimulus observed during successive reversals. The difficulty in performing reversals further demonstrates a well-developed memory associated with a resistance to forgetting even after long intervals of time.

According to the same types of criteria, the visual capacities of *Phaner* are more highly developed. Nonspatial visual cues are strongly attended to, even when irrelevant to obtaining reward. In *Phaner,* individual object discriminations are rapidly learned and equal the rates of learning observed in certain diurnal primates (Rumbaugh and Gill, 1972; Cooper, 1978). Although during Transfer Index, acquisition of individual problems occurs in fewer trials than for *Microcebus,* reversal performance is not better developed. Therefore it is merely the discriminative ability which is more acute in *Phaner,* in relation to the higher degree of visual attention and responsiveness to the discrete nonspatial visual cues. The more rapid reversal improvement by *Phaner* as compared to *Microcebus* during visual SDR (Fig. 1), in contrast to the similar reversal performance during TI cannot be explained only by differences in attention and/or discrimination ability. The form of the intraproblem learning curves, and especially the results of reversals (or nonreversals) after long periods of time, suggests that forgetting is responsible for this difference. In the case of successive visual reversals involving the same stimuli, interference from previous learning can cause forgetting (proactive interference, Mackintosh, 1974). However, this type of interference cannot develop during reversal of TI problems, since each individual discrimination consists of an independent pair of stimuli.

The general difficulty of prosimians in performing visual reversals has previously been remarked by other authors. Stevens (1965) attributed this difficulty to stimulus perseveration, a tendency which was also present during learning set formation in *Lemur* (Cooper, 1973). This tendency was also found in a study of reversal learning in *Lemur* (Rumbaugh and Arnold, 1971), and in a subsequent study of extinction (Arnold and Rumbaugh, 1971). Finally, the retention of individual visual discriminations over long periods of time has been shown in other nocturnal prosimians (Erlich and Musicant, 1976). The present study supports the importance of stimulus perseverance and memory in nocturnal prosimians, and further suggests differential species capacities for forgetting visually learned associations.

How can these species learning differences be related to other species typical distinctions? Since both species are nocturnal and sympatric, it might be assumed that they are subject to similar environments in terms of the quantity and quality of light available. However, because most of the activities of each species take place at a different range of heights in the forest (Chapter 2, Fig. 11) light differs in composition and intensity in the respective micro-environments due to differential absorbtion by the biotope (see Chapter 5 for discussion of light in the forest and in the canopy). In addition, *Phaner* leaves the nest during twilight, whereas *Microcebus* begins activity later in the night during lower light levels (Chapter 7). Distinct differences also exist in the utilization of food and other resources, social organization, and communication (Chapters 2 and 3).

Microcebus and *Phaner* differ with respect to the structure of the retina and the eye, which partly explains certain differences in visual capacities. The total visual field and the degree of binocular overlap is greater in *Microcebus* than in *Phaner* (Pariente, 1978). In the retina the area centralis of *Phaner* is more conspicuous, and the ratio of receptor to ganglion cells lower than in *Microcebus* (Pariente, 1976; Cooper *et al.,* 1979). These features indicate a greater reliance on central as compared to peripheral vision in *Phaner,* as well as a higher degree of visual acuity (necessary for fine object discrimination). Comparatively, for *Microcebus* there is a relative loss in acuity, but nevertheless a gain in sensitivity (associated with lower light levels in the forest interior), and in the importance of peripheral vision (advantageous for the detection of flying insects).

Phaner feeds primarily on gums, which are localized sources of nutrients, especially important during the winter season. Gums are mainly found in *Terminalia* spp. trees at precise locations in the home range. Field observations suggest that vision plays an important role in learning and memorizing itineraries between food sources. *Microcebus* is certainly also capable of memorizing paths between food sources. However, the preferred food of *Microcebus* is insects, which are often caught while flying by means of a species typical grasping behavior, in which both vision and audition are coordinated for detection and capture. In this case, movement of the stimulus is an important parameter, and careful visual inspection would be disadvantageous to prey capture. Even in the test situation, *Microcebus* persisted in rapid response following decision from a distance.

The difference in preference for mobile and nonmobile food resources explains the qualitative differences in visual attention between the two species (Pariente, 1978). Indeed, from observations in captivity of *Microcebus,* small insects are not detected if immobile and silent, whereas *Phaner* quickly locates any novel object in the home cage.

The results demonstrate that both species possess well-developed capacities for memorization, accentuated by poor abilities for forgetting. This tendency is especially pronounced in *Microcebus,* for which there may be no obvious ecological pressure for forgetting to evolve. For *Phaner* there exists a need to find new sites of vegetal exudates when old sources of gums are exhausted, and thus frequently readjust itineraries at irregular intervals. Adaptation to exploit these gums, which are relatively less accessible to other species, would include not only specializations of the claws for climbing large smooth trunks, and of the teeth for scraping (Chapter 3), but also capacities associated with vision for detecting and memorizing complex itineraries.

Social organization and communication contribute additional selection pressures influencing visual capacities in terms of attention to, discrimination, and recognition of conspecifics. Whereas *Microcebus* is solitary during activity (Martin, 1973) and relies primarily on deferred olfactive and auditory communication at a distance (Charles-Dominique, 1978), social communication in *Phaner* is based almost exclusively on visual and auditory exchanges (Chapter 3). In association with visual communication and recognition, only *Phaner* possesses distinct markings of the pelage which are easily distinguishable by conspecifics in natural light conditions (Pariente, 1976). Because of these and other characteristics, the social organization of *Phaner* has been described by Charles-Dominique (1978) as "pregregarious," and is unusual for a nocturnal primate.

This type of social system and the associated visual learning capacities may represent a transitional state from a nocturnal solitary organization to the gregarious social system typical of diurnal species.

REFERENCES

Arnold, R. C., and Rumbaugh, D. M. (1971). Extinction: A comparative study of *Lemur* and *Cercopithicus*. *Folia Primatol.* **14**, 161–170.

Bitterman, M. E. (1965). The evolution of intelligence. *Sci. Am.* **212** (1), 92–100.

Bolles, R. C. (1970). Species-specific defence reactions and avoidance learning. *Psychol. Rev.* **77**, 32–48.

Charles-Dominique, P. (1978). Solitary and gregarious prosimians: Evolution of social structures in Primates. *In* "Recent Advances in Primatology" (D. J. Chivers and K. A. Joysey, eds.), Vol. 3, pp. 139–149. Academic Press, New York.

Charles-Dominique, P., and Martin, R. D. (1970). Evolution of lorises and lemurs. *Nature (London)* **277**, 257–260.

Cooper, H. M. (1973). Learning set formation in *Lemur fulvus abbifrons* and *Hapalemur griseus griseus*. Thesis, Florida State University, Tallahassee.

Cooper, H. M. (1978). Learning in Prosimians. *In* "Recent Advances in Primatology" (D. J. Chivers and J. Herbert, eds.), Vol. 1, pp. 941–944. Academic Press, New York.

Cooper, H. M., Kennedy, H., Magnin, M., and Vital-Durand, F. (1979). Thalamic projections to area 17 in a prosimian primate *Microcebus murinus*. *J. Comp. Neurol.* **187**, 145–168.

Erlich, A., and Musicant, A. D. (1976). Visual discrimination learning and memory in nocturnal prosimians. *Anim. Learn. Behav.* **4** (4), 431–435.

Gellermann, L. W. (1933). Chance orders of alternating stimuli in visual discrimination experiments. *J. Genet. Psychol.* **42**, 207–208.

Kay, R. F. (1975). The functional adaptations of primate molar teeth. *J. Phys. Anthropol.* **43** (2), 195–215.

Mackintosh, N. J. (1974). "The Psychology of Animal Learning." Academic Press, New York.

Martin, R. D. (1973). A review of the behaviour and ecology of the Lesser Mouse Lemur (*Microcebus murinus*, J. F. Miller 1777). *In* "Comparative Ecology and Behaviour of Primates" (R. P. Michael and J. H. Crook, eds.), pp. 1–68. Academic Press, New York.

Masterton, B., Skeen, L. C., and Robards, M. J. (1974). Origins of anthropoid intelligence. II. Pulvinar extrastriate system and visual reversal learning. *Brain, Behav. Evol.* **10**, 322–353.

Pariente, G. F. (1976). Etude eco-physiologique de la vision chez les Prosimians Malgache. Thesis, Université des Sciences et Techniques du Languedoc.

Pariente, G. F. (1978). The role of vision in Prosimian behavior. *In* "The Study of Prosimian Behavior" (G. A. Doyle and R. D. Martin, eds.), pp. 411–459. Academic Press, New York.

Riddell, W., Corl, K., Bennett, V. D., and Reimers, R. O. (1974). Discrimination learning differences and similarities as a function of brain index. *Physiol. Behav.* **13**, 401–405.

Riddell, W. I., Corl, K., and Gravetter, F. (1976). Cross order comparisons using indexes of cerebral development. *Bull, Psychon. Soc.* **8** (1), 578–580.

Rumbaugh, D. M. (1969). The transfer index, an alternative measure of learning set. *Proc. Int. Congr. Primatol.*, 2nd, *1968* Vol. 1, pp. 267–273.

Rumbaugh, D. M., and Arnold, R. C. (1971). Learning: A comparative study of *Lemur* and *Cercopithicus*. *Folia Primatol.* **14**, 154–160.

Rumbaugh, D. M., and Gill, T. V. (1972). The learning skills of Great Apes. *J. Hum. Evol.* **2**, 171–179.

Rumbaugh, D. M., Bell, C. L., and Gill, T. V. (1972). Two discrimination test apparatuses for Primates. *Behav. Res. Methods Instrum.* **4** (1), 6–10.

Seligman, M. E. P. (1970). On the generality of the laws of learning. *Psychol. Rev.* **77**, 406–418.

Stevens, D. A. (1965). A comparison of learning in rhesus monkeys, cebus monkeys, lemurs, and Burmese cats. Unpublished Ph.D. thesis, University of Oregon, Eugene.

Szalay, F. S., and Seligsohn, D. (1977). Why did the strepsirhine tooth comb evolve? *Folia Primatol.* **27**, 75–82.

Warren, J. M. (1973). Learning in vertebrates. *In* "Comparative Psychology, a Modern Survey" (D. A. Dewsbury and D. A. Rethlingshafer, eds.), pp. 471–509. McGraw-Hill, New York.

Summaries of Chapters

1. The climate of the West coast of Madagascar is characterized by a 7- to 8-month dry season. In the closed deciduous forest adapted to these conditions, about 200 woody plant species have been collected among which evergreen trees and lianas are present in limited number. Leafing times have been determined from the analysis of litter fall completed by records of herbarium specimens. A classification of leafing patterns is proposed, and flowering and fruiting patterns are briefly described. The seasonal variations of potential foods available to prosimians can be estimated from these data. The possible importance of toxic compounds has also been considered according to the results of a screening survey of alkaloids in leaves. The percentage of plants likely to be toxic (15%) has the same order of magnitude as in other tropical forests, but positive results appear more frequently among the evergreen species which are subject to folivore pressure during the dry season.

2. Annual variations in food production and composition in the forest of the west coast of Madagascar are presented in relation to different feeding strategies of five sympatric species of nocturnal prosimians. These strategies involve combination of physiological and behavioral characteristics which have been analyzed in the laboratory, and are presented in separate chapters in this volume. The results of laboratory tests are interpreted according to comparative results of the field studies. In *Cheirogaleus medius* and, to a lesser extent, in *Microcebus murinus,* the annual cycle of activity and lethargy plays a major role in the adaptation to variations in food availability. Nevertheless, seasonal variations in the patterns of food choices are a necessary complement to cope with overall changes in composition of the forest resources. In *Microcebus coquereli* and *Phaner furcifer,* behavioral specializations on insect secretions and plant exudates, which are the basis of their respective diets, are also complemented by additional cycles involving physiological variations. The most folivorous species, *Lepilemur ruficaudatus,* is also subject to variations in physiological/behavioral patterns, following the environmental changes in climate and plant production.

3. The ecology and social relations of *Phaner furcifer* were studied in 1973–1974. This species is specialized on a diet mainly composed of gums and other tree exudates, and population density appears to depend on the distribution of gum producing *Terminalia* spp. trees. Social communication is mainly based on vocal signals which are described, while olfaction does not seem to play an important role as in other nocturnal prosimians. Within one sex, territories are adjacent with small zones of overlap. The territory of one male may overlap with these of several females, and there are permanent social bonds between these individuals. Males gather in "meeting areas" and engage in mutual vocalization. This type of society has been refered to as "pregregarious."

4. *Microcebus coquereli* was observed during the dry season (austral winter). Twenty-one individuals were followed using techniques of radiotracking and/or direct behavioral observation. The diet is based on the peculiar exploitation of secretions of homopteran larvae colonies. Nevertheless, population spatial distribution mainly depends on insect food resources. Each individual builds and utilizes up to 12 spherical nests which are located in the central area of the home range. The central areas of males are separated and overlap the intercalated central zones of female home ranges. Social and "individual" activities are both temporally and spatially distributed between the central and peripheral home range areas. The adaptive value of these behavioral patterns is discussed in terms of utilization of food resources.

5. The spectral energy distribution of light was studied inside and outside the forest of the west coast of Madagascar, both during the day and at night. Comparisons with measures made in other forests of Madagascar show that there is a relative homogeneity in open day light, but the vegetation may differently affect the energy distribution. Nocturnal light in the forest is characterized by large variations in intensity and energy distribution according to time, season, and meteorological conditions. The efficiency of nocturnal prosimian vision is discussed in terms of contrast cues produced by directly illuminated objects against the background forest light.

6. The five species of nocturnal malagasy prosimian primates were studied for 3 years in simulated climatic conditions. Observations concerned body weight, food intake, duration of the reproductive period, and variations in locomotor activity. In all five species evident annual rhythms were recorded, and the reproductive cycle was synchronized by variations in the photoperiod. However, only *Microcebus murinus* and *Cheirogaleus medius* undergo a noticeable variation in body weight (up to 20% deviation from the annual mean). These two species demonstrated a very distinct physiological cycle including an increase in activity during the summer months, followed by increased food intake and rapid weight gain before winter "lethargy." Hypothermia was found only in these two species. The results are interpreted in terms of specific adaptive strategies in response to seasonal variations in the natural habitat.

7. Circadian rhythms in locomotion and feeding activities were recorded during 2 years in the five species of nocturnal prosimians. Activity was relatively reduced in winter compared to that in summer, and important peaks of activity occurred during periods of estrus. The onset of activity was very regular and occurred shortly after dusk throughout the year, whereas cessation of activity was more variable and occurred long before dawn in most species. All species showed characteristic circadian rhythms which varied throughout the year, depending on the species. These were compared to annual variations of the natural environment and interpreted in terms of specific adaptation.

8. A stable diet was provided to the five different prosimian species in captivity over a period of 2 years. Each species demonstrated a typical pattern of food choice. Annual variations in food intake were obvious in two Cheirogaleinae (*Microcebus murinus* and *Cheirogaleus medius*). Variations in dietary composition may result from the different seasonal food choices of *Microcebus coquereli, Phaner furcifer,* and *Lepilemur ruficaudatus.* A multivariable analysis helps to understand how the different behavioral patterns serve to maintain specific strategies observed in the field.

9. The fecal marking of *Cheirogaleus medius* is characterized by a definite annual periodicity. Marking is maximum at the end of the summer resulting in a monthly mean of feces greatly exceeding the annual mean. During the active summer period, the quantity of fecal marks deposited by the animal is not directly proportionnal to food intake. The ratios of the weight of sample fecal marks to either locomotor activity, the dry weight of food consumed, or to nonassimilable food ingested are used to define marking indices. These indices help to understand the seasonal variation of marking behavior and its particular physiological and behavioral significance.

10. Learning capacities were studied in the two species of nocturnal prosimians, *Microcebus murinus* and *Phaner furcifer.* Visual learning and behaviors associated with vision were quantitatively and qualitatively different. The differences concerned attention to spatial and nonspatial visual cues, discrimination abilities, reversal performance, memory, and forgetting. The visual capacities of each species are correlated with specific anatomical and ethoecological characters.

Index